Acknowledgments

I wish to express my sincere appreciation and gratitude to my supervisors, Prof. Dr. Mohamed Raafat Mahmoud El-Koussy and Prof. Dr. Iman Salah El Deen Elmahalawi, for their valuable guidance, support, comments and inspiration during my PhD candidature. Thanks are also extended to the members of Mechanical Testing Lab of Metallurgical Department at Cairo University for their help, with special thanks for Mr. Mohamed Mamdoh for his extensive effort and support in all testing stages.

I also wish to deeply thank the Project Services Company (PSC), specially the Managing Director of PSC, Mr. Mohamed Kandil, for offering the materials and facilities available at their plant, without which this work would have not been possible.

I can't forget PSC Quality and Production members for assisting in the fabrication of stress rupture machine and in the welding and preparation of tested specimens.

Last but not least, my warm appreciation and respect is due to my beloved wife for allowing me to stay away from home, taking care of our children, her understanding, continuous encouragement and patience throughout my PhD candidature and which made all this possible.

Dedication

I wish to dedicate this thesis to my mother, my father, my beloved wife and my dear children.

Table of Contents

ACKNOWLEDGEMENTS	I
DEDICATION	II
TABLE OF CONTENTS	III
LIST OF TABLES	VII
LIST OF FIGURES	XI
NOMENCLATURE	XVI
ABSTRACT	XVII
CHAPTER 1: INTRODUCTION	1
CHAPTER 2 : LITERATURE REVIEW	4
2.1. MATERIALS FOR POWER GENERATION PLANTS-OVER VIEW	4
2.1.1. CARBON-MANGANIZ STEELS (C-Mn STEELS)	7
2.1.2. Mo-STEELS	8
2.1.3. Cr-Mo STEELS	9
2.1.4. 9- 12% Cr STEELS	10
2.1.5. HIGH PERFORMANCE CREEP RESISTANT STEELS FOR 21st CENTURY POWER PLANTS	12
2.1.6. AUSTENITIC STAINLESS STEELS	12
2.2. ALLOY DESIGN FOR 9-12% Cr STEELS	15
2.2.1. CHROMIUM (Cr)	15
2.2.2. MOLYBDENUM, TUNGSTEN AND RHENIUM (Mo, W & Re)	16
2.2.3. VANADIUM, NIOBIUM AND TITANUM (V, Nb & Ti)	16
2.2.4. CARBON AND NITROGEN (C & N)	17
2.2.5. BORON (B)	17
2.2.6. MANGANESE AND SILICON (Mn & Si)	18
2.2.7. NICKEL, COPPER AND COBALT (Ni, Cu & Co)	18

2. 2. 8. COMBINATION EFFECT OF ALLOYING ELEMENTS IN 9-12% Cr STEELS	18
2. 3. MICROSTRUCTURE OF 9-12% Cr CREEP RESISTANT STEELS	22
2. 3. 1. COMPARISON OF THE MICROSTRUCTURE OF P91, P92 AND E922	24
2. 4. MICROSTRUCTURE STABILITY AND AGEING EFFECT	26
2. 5. UNDERSTANDING OF MAIN CHARACTERISTICS OF CREEP RESISTANT STEELS:	30
2. 5. 1. CREEP AND STRESS RUPTURE STRENGTH	30
2. 5. 1. 1. PREDICTION OF LONG-TERM CREEP BEHAVIOR	34
2. 5. 1. 1. 1. EXTRAPOLATION SCHEMES	35
2. 5. 2 TECHNIQUES OF CREEP TESTING AND ASSESSMENT OF CREEP STRENGTH	37
2. 5. 2. 1. CONVENTIONAL TENSILE CREEP TESTING	39
2. 5. 2. 2. HYPERBOLIC-WEIGHT CONSTANT-STRESS METHOD	39
2. 5. 2. 3. BALANCED BEAM WITH CAMS	40
2. 5. 2. 4. CONSTANT STRESS TEST FOR USE WITH LOW FORCES	41
2. 5. 2. 5. HIGH-TEMPERATURE CONSTANT-STRESS COMPRESSION CREEP TESTING	41
2. 5. 2. 6. SMALL-PUNCH CREEP TEST	42
2. 6. CREEP DAMAGE AND CREEP FRACTURE	43
2. 6. 1. CREEP MECHANISMS OF METALS	44
2. 6. 1. 1. DISLOCATION CREEP (GIVING POWER-LAW CREEP)	44
2. 6. 1. 2. DIFFUSION CREEP (GIVING LINEAR-VISCOUS CREEP)	46
2. 7. CHRACTRISTICS OF 9-12% Cr CREEP RESISTANT STEELS	47
2. 7. 1. LONG-TERM TESTING	48
2. 7. 2. THERMAL FATIGUE RESISTANCE	49
2. 7 .3. STEAM-SIDE OXIDATION RESISTANCE	50
2. 8. HEAT TREATMENT AND MICROSTRUCTURE OF 9-12% Cr STEELS	52
2. 8. 1. CONTINUOUS COOLING DIGRAMS AND TRANSFORMATION BEHAVIOR	53
2. 9. WELDABILITY AND WELDING OF 9-12% Cr STEELS	55

2. 9 .1. AVOIDANCE OF HYDROGEN INDUCED CRACKING	56
2. 9 .1. 1. CALCULATION OF PREHEATING TEMPERATURE	57
2. 9. 2. HEAT TREATMENTS FOR CrMo STEELS AND WELDED JOINTS	57
2. 9. 2. 1. INTERMEDIATE HEAT TREATMENT	58
2. 9. 2. 2. TEMPER EMBRITTELMENT	58
2. 9. 3. WELDING AND WELDING CONSUMABLES FOR CrMo STEELS	59
2. 9. 4. CHARACTERISTIC OF WELDED JOINT	59
2. 9. 4. 1. PARENT METAL (PM)	60
2. 9. 4. 2. HEAT AFFECTED ZONE (HAZ)	60
2. 9. 4. 2. 1. GRAIN-REFINED HEAT AFFECTED ZONE (GRHAZ)	61
2. 9. 4. 2. 2. GRAIN-COARESEND HEAT AFFECTED ZONE (GCHAZ)	61
2. 9. 4. 2. 3. INTERCRETICAL HEAT AFFECTED ZONE (ICHAZ)	61
2. 9. 4. 3. WELD METAL (WM)	61
2. 9. 4. 4. DISSIMILAR WELDING OF 9-12% Cr STEELS	61
2. 9. 5. FAILUR MECHANISMS OF WELDED JOINTS IN THE CREEP RANGE	62
2. 9. 5. 1. FERRITIC STEELS – CREEP RUPTURE	62
2. 9. 5. 2. DESIGNATION OF CRACK LOCATION (TYPES OF CRACKING)- QUICK OVER VIEW	64
2. 9. 6. FACTORS AFFECTING RESISTANCE TO CREEP FRACTURE	66
CHAPTER 3: EXPERIMINTAL WORK	68
3. 1. EXPERIMINTAL PROCEDURE	68
3. 1. 1. MATERIALS	69
3. 1. 2. APPLIED PREHEAT AND HEAT TREATMENT	74
3.1.2.1. PREHEAT AND INTERMEDIATE POST WELD HEAT TREATMENT	74
3.1.2.2. HEAT TREATMENT	75
3. 1. 3. SAMPLES TRACEABILITY AND CODING SYSTEM	76
3. 1. 4. APPLIED TESTING METHODS	77
3. 1. 4. 1. TENSILE TEST	77
3. 1. 4. 2. IMPACT TEST	79
3. 1. 4. 3. HARDNESS TEST	79
3. 1. 4. 4. STRESS RUPTURE TEST	79

3. 1. 4. 5. STRESS CORROSION CRACKING TEST	81
3. 1. 4. 6. MICROSTRUCTURE INVESTIGATION	82
3. 1. 4. 6. 1 OPTICAL MICROSCOPE	82
3. 1. 4. 6. 2 SCANNING ELECTRON MICROSCOPIC (SEM) AND ENERGY DISPERSIVE X-RAY (EDS)	83
3. 1. 4. 6. 2. 1. ROLES OF CARBIDES IDENTIFICATIONS	84
3. 1. 4. 6. 3. X-RAY DIFFRACTION	86
3. 1. 4. 6. 4. FERRITESCOPE EXAMINATION	87
CHAPTER 4: RESULTS AND DISCUSSION	**88**
4. 1. MECHANICAL PROPERTIES	88
4. 1.1. HARDNESS TESTING RESULTS	88
4. 1. 2. TENSILE TEST RESULTS	91
4. 1.3. IMPACT TEST RESULTS	94
4. 1.4. STRESS RUPTURE TESTS	96
4.1.4.1. EFFECT OF WELDING AND PWHT CONDITIONS ON TIME TO RUPTURE	96
4. 1. 5. STRESS CORROSION CRACKING RESULTS	108
4. 2. MICROSTRUCTURE	112
4. 2. 1. BASE METAL	112
4. 2. 2. WELD METAL AND HAZ	115
CHAPTER 5: SUMMARY AND CONCLUSION	**141**
REFERENCES	**142**

List of Tables

Table 2.1: Pressure vessel Steels for high-Temperature Service	6
Table 2.2: Chemical compositions of creep resistant ferritic steels for power plants	7
Table 2.3: Nominal Chemical Compositions of Austenitic Steels for Boiler (wt%)	13
Table 2.4: Details of high chromium steels, chemical compositions in wt%	25
Table 2.5: Dislocation density & mean sub-grain size of as received P91, P92, E911 steels	26
Table 2.6: Initial microstructure of Heat treated (Normalized in the range 1040-1090 °C and Tempered in the range of 740-780 °C) P91, P92 and E911 Steels.	27
Table 2.7 : Equivalent Conditions Based on Larson–Miller Parameter.	36
Table 2.8: Precipitations that can be found in creep resistant CrMo steels.	47
Table 2.9 : Statistics of ECCC databases for steels P92, E911 and P91.	48
Table 3.1: Summary of the experimental work program presented through this research.	68
Table 3.2: Chemical Composition for the as received P91 steel compared to material standard, ASME SA355.	69
Table 3.3: Mechanical properties and microstructure of the as received investigated steel P91.	69
Table 3.4: Chemical composition of GTAW rod ER90S-B9 and SMAW electrode E9015-B9 used during welding, wt%.	70
Table 3.5: Summary of welding procedures used during pipes welding.	71
Table 3.6: Coding system for different sample condition (Heat Input/ Heat Treatment).	77
Table 3.7: Chemical composition of Villella's Reagent.	83
Table 3.8: XRD peaks analysis from the XRD pattern of the as quenched specimen (Standard Martensitic sample).	87
Table 4.1: Time to rupture under a stress of 135 MPa at different temperatures for different heat input and heat treatment conditions.	96
Table 4.2: Stress rupture test results for Low heat input condition followed by PWHT at 760 °C for 3.5 hrs (L760).	102

Table 4.3: Stress rupture test results for Medium heat input condition followed by PWHT at 760 °C for 3.5 hrs (M760).   103

Table 4.4: Stress rupture test results for High heat input condition followed by PWHT at 760 °C for 3.5 hrs (H760)   104

Table 4.5: Stress rupture test results for Low heat input condition followed by normalizing at 1050 °C for 0.5 hr and tempering at 760 °C for 3.5 hrs (L1050):   105

Table 4.6: Stress rupture test results for Medium heat input condition followed by normalizing at 1050 °C for 0.5 hr and tempering at 760 °C for 3.5 hrs (M1050)   106

Table 4.7: Stress rupture test results for High heat input condition followed by normalizing at 1050 °C for 0.5 hr and tempering at 760 °C for 3.5 hrs (H1050)   107

Table 4.8: Results of EDS analysis related to precipitates indicated through Fig.4.21, wt %.   114

Table 4.9: Peaks analysis of X-Ray diffraction pattern of P91 base metal after different treatment conditions.   115

Table 4.10: Ferrite content measurement within Weld metals, HAZ and Base metals of different heat inputs/Heat Treatment conditions.   120

Table 4.11: Results of EDS analysis related to precipitates indicated through Fig.4.44, wt %.   132

Table 4.12: Results of EDS analysis related to precipitates indicated through Fig.4.45, wt %.   133

Table 4.13: Results of EDS analysis related to precipitates indicated through Fig.4.46, wt %.   134

Table 4.14: Results of EDS analysis related to precipitates indicated through Fig.4.47, wt %.   135

List of Figures

Fig. 2.1: Temperature dependence of tensile strength for selected metallic materials 4

Fig. 2.2: The variation in wall thickness with material grade (with an internal diameter of 200 mm) for service at 593.5 °C & 320 bar. 5

Fig. 2.3: Strength properties of CMn- and Mo- steels. 8

Fig. 2.4: Strength properties of CrMo- steels. 9

Fig. 2.5: Strength properties of 9-12% Cr- steels. 10

Fig. 2.6: Evolution of ferritic steels for boilers 11

Fig. 2.7: Effect of (Ti1Nb)/C ratio on creep rupture strength of 18Cr10NiNbTi steel..... 14

Fig. 2.8: Effect of Cu content on creep rupture strength of 18Cr9NiCuNbN steel. 14

Fig. 2.9: Evolution of Austenitic steels for boilers 15

Fig. 2.10: Cr-Fe equilibrium diagram 16

Fig. 2.11: Effect of alloying element additions on hardness after nitriding 17

Fig. 2.12: Effect of Mo+W and V+Nb on creep rupture strength of 12%Cr steels 19

Fig.2.13: Effect of Mo and W contents on 10^5 h creep rupture strength at 650°C and Charpy absorbed energy at 20°C of 12%Cr turbine steels (0.13C0.05Si0.5Mn11.2Cr-0.8Ni 0.2V0.05Nb0.05N–Mo–W). 19

Fig. 2.14: Effect of W1Mo and Co contents on creep rupture strength of 12%Cr turbine steels. 20

Fig. 2.15: General concept of alloy design for heat resistant steels 21

Fig. 2.16: Alloy design of 12Cr0.4Mo2WCuVNb steel................... 22

Fig. 2.17: Typical microstructure of tempered martenstitc 9–12% Cr steel................... 23

Fig. 2.18: Schematic illustration of microstructures of ferritic steels................... 24

Fig. 2.19: Schematic illustration of microstructures of austenitic steels 25

Fig. 2.20: Microstructures of the as received steels (a) P91, (b) P92 and (c) E911 showing the elongated sub-grains of tempered martensite (TEM) 26

Fig. 2.21: Coarsening of particles during aging 28

Fig. 2.22: Z-Phase Particle in a 11CrWCoVNbNB Steel 29

Fig. 2.23: Stages of plastic strain as a function of time during creep testing at temperatures above 0.5 ^ T. The corresponding creep strain rate is plotted above......... 31

Fig. 2.24: The variation of creep rate with stress and temperature 31

Fig. 2.25: Illustration of creep and stress rupture ……………………………………... 32

Fig. 2.26 : Creep strain versus time test results in a 0.5 wt% Mo, 0.23 wt% V steel (A) Constant stress, variable temperature. (B) Constant temperature, variable stress …... 33

Fig. 2.27: Stress rupture plots for various temperatures …………………………….. 34

Fig. 2.28: Summary of Larson-Miller relation ……………………………………….. 36

Fig. 2.29: Summary of Sherby-Dorn relation ……………………………………….. 37

Fig. 2.30: Comparison of creep and stress rupture tests …………………………….. 38

Fig. 2.31: A chart illustrating the most used methods for creep testing …………….. 38

Fig. 2.32: Typical creep test set-up ……………………………………………….. 39

Fig. 2.33: Hyperbolic-weight constant-stress apparatus …………………………….. 40

Fig. 2.34: Balanced beam with cams for creep testing ……………………………….. 40

Fig. 2.35: Constant stress testing system for use with low forces …………………… 41

Fig. 2.36: Knife-edge configuration for constant-stress compression creep testing ……… 42

Fig. 2.37: Schematic illustration of the small punch die ……………………………... 43

Fig. 2.38: Creep damage in materials ……………………………………………….. 44

Fig. 2.39: The climb force on a dislocation ………………………………………….. 45

Fig. 2.40: How diffusion leads to climb ……………………………………………... 45

Fig. 2.41: How the climb-glide sequence leads to creep……………………………... 46

Fig. 2.42: How creep takes place by diffusion ……………………………………….. 47

Fig. 2.43: Thermal Conductivity of Some Current and Proposed Header and Steam Separator Materials ……………………………………………………………………. 50

Fig. 2.44: Mean Coefficient of Linear Expansion for some Current and Proposed Header and Steam Separator Materials ………………………………………………... 50

Fig. 2.45: Scale microstructure of P91 exposed for 1 000 h at 650°C in Ar-50% H_2O….. 51

Fig. 2.46 : Mass change data for Cr steels exposed in Ar-50% H2O and in air at 650°C 52

Fig. 2.47. a.: CCT diagram of Grade 91	53
Fig. 2.47. b.: CCT diagram of Grade 911	53
Fig. 2.48: Samples of (a) Grade 91 after 1070°C+780°C and (b) Grade 911 after 1060°C+780°C	54
Fig. 2.49: Delta-ferrite fraction in 911 steel as function of temperature. COST experimentation (30 min holding time in furnace); and CSM experimentation (120 min holding time).	54
Fig. 2.50.: The fraction of martensite in the microstructure against the temperature below Ms.	55
Fig. 2.51: Temperature cycle and heat control during welding and PWHT of martensitic steel P91, E911 and P92.	58
Fig. 2.52. Weldment different regions ………………………………………………..	60
Fig. 2.53: Weld strength factors for 100.000 h creep rupture strength ………………….	63
Fig.2.54: Classification of crack locations ………………………………………………	65
Fig.2.55: Schematic view of damage progress ………………………………………...	65
Fig. 3.1: Portable heat treatment machine with ceramic heaters used for preheating and intermediate post weld heat treatment.	74
Fig. 3.2: Intermediate post weld heat treatment applied on each pipe after welding …….	74
Fig. 3.3: Water/Oil cooling saw cutting of welded pipes – First step of samples preparation.	75
Fig. 3.4: Different post weld heat treatments applied on a part of each pipe after welding. (a) Tempering. (b) Normalizing and Tempering.	76
Fig. 3.5: Oven used for tempering and normalizing/tempering heat treatments processes with heating/cooling rates control.	76
Fig. 3.6: Cross weld tensile specimen; all dimensions are in mm ……………………….	78
Fig. 3.7: All weld metal tensile specimen; all dimensions are in mm ………………..	78
Fig. 3.8: Direction and location of V-notch of the impact specimen ………………..	79
Fig. 3.9: Cross weld stress rupture specimen; all dimensions are in mm ………………..	80
Fig. 3.10: Creep test machine used in the stress ruptures testing……………………..	80
Fig. 3.11: Stress-corrosion test using U-bend specimens...	81
Fig. 3.12: JEOL JSM-5410 - Scanning electron microscopic (SEM) with energy dispersive X-ray spectrometer; used for both SEM and EDS analysis.	83
Fig. 3.13: Characteristic X-ray Spectra from various carbides ………………………….	84
Fig. 3.14: characteristic X-ray spectra from: (a) Laves Phase. (b) from VC. …………	85
Fig. 3.15: illustrates the types of carbides identified and the distribution of major alloying elements in them **[83].**	85

Fig. 3.16: X-Ray diffraction pattern of P91 base metal normalized at 1050 °C for 0.5 hr and quenched in water.   87

Fig. 4.1. Hardness profiles along weldements for various heat input conditions: (a) As-welded condition. (b) Subcritical PWHT. (c) N&T heat treatment.   89

Fig. 4.2. (a) Carbone equivalent measurement along weldements of different heat inputs in the subcritical PWHT conditions; (b) Carbon activity, a_c, of the different steels included P91 at different temperatures.   90

Fig.4.3. Carbone equivalent measurement along weldement of medium heat input after normalizing and tempering.   91

Fig.4.4. Ultimate tensile strength of different heat treatment/heat input conditions. (a) Weld Metal. (b) HAZ.   92

Fig. 4.5. Reduction in area of different heat input/heat treatment conditions. (a) Weld metal. (b) HAZ.   93

Fig.4.6. Impact toughness of different heat input/heat treatment conditions. (a) Weld metal. (b) HAZ. (c) Base Metal.   95

Fig. 4.7: Time to rupture results of different heat input conditions at 135 MPa and Temperature 631 °C.   97

Fig. 4.8: Stress rupture test results of different heat input conditions of Sub-critical PWHT.   97

Fig.4.9.: Stress rupture test results of different heat input conditions of Normalized/tempered.   98

Fig. 4.10.: Creep rate resulted from stress rupture test of different heat input and Heat treatment conditions.   98

Fig. 4.11: Elongation and Reduction of Area obtained from Stress rupture tests   100

Fig.4.12: Elongation of Low and High heat input in both sub-critical and normalizing/tempering PWHT conditions   101

Fig.4.13: Reduction in area of Low and High heat input in both sub-critical and normalizing/tempering PWHT conditions   101

Fig 4.14: Fracture appearance and location for Low heat input condition followed by PWHT at 760 °C for 3.5 hrs (L760).   102

Fig 4.15: Fracture appearance and location for medium heat input condition followed by PWHT at 760 °C for 3.5 hrs (M760).   103

Fig 4.16: Fracture appearance and location for High heat input condition followed by PWHT at 760 °C for 3.5 hrs (H760). — 104

Fig 4.17: Fracture appearance and location for Low heat input condition followed by normalizing at 1050 °C for 0.5 hr and tempering at 760 °C for 3.5 hrs (L1050). — 105

Fig 4.18: Fracture appearance and location for medium heat input condition followed by normalizing at 1050 °C for 0.5 hr and tempering at 760 °C for 3.5 hrs (M1050) — 106

Fig 4.19: Fracture appearance and location for High heat input condition followed by normalizing at 1050 °C for 0.5 hr and tempering at 760 °C for 3.5 hrs (H1050) — 107

Fig. 4.20: Weld area of L760, M760 and H760 tested for 0.5 day. Percentage of pitting in the weld area increased with increase in the heat input during welding for the tempered conditions. Low heat input condition showed no pitting. — 108

Fig. 4.21: Weld area of H1050, M 1050 and L1050 tested for 0.5 day. Percentage of pitting in the weld area increased with increase in the heat input during welding for the normalized/tempered conditions. Low heat input condition showed no pitting. — 109

Fig 4.22: Weld area of L760, M760 and H760 tested for 5, 7.5 and 4 days respectively — 110

Fig. 4.23: Weld area of H1050, M 1050 and L1050 tested for 7, 4 and 5.5 days respectively. Percentage of corrosion in the weld and HAZ areas increased with increase in the heat input during welding for the normalized/tempered conditions. — 111

Fig. 4.24: SEM Microstructure of P91 Base Metal: (a) As Received Base Metal, **91 %** measured ferrite **(F).** (b) Sub-critical PWHT Base Metal, **95% F.** (c) Normalized/Tempered Base Metal, **93% F**. — 113

Fig. 4.25: Location of precipitates subjected to EDS in the As received Base Metal. (a) Along Martensite lathes. (b) Along Ferrite grains. — 113

Fig. 4.26: X-Ray diffraction pattern of P91 base metal after different treatment conditions: (a) B350. (b) B760. (c) B1050. — 114

Fig.4.27: Schematic representations of microstructures developed in weld metal and HAZ as function of peak temperature during welding **[94]** and CCT diagram of steel P91. — 116

Fig. 4.28: Microstructure of weld metals of as-welded condition for different heat inputs showing fine un-tempered Martensite and acicular ferrite: a) L350, 75% ferrite (F); b) M350, 77.5% F; c) H350, 82% F. — 117

Fig. 4.29: SEM Microstructure of As-Welded condition Weld Metals: (a-b) L350. (c-d) M350. (e) H350.   118

Fig. 4.30: Martensite microstructure at different heat affected areas between weld passes as a result of re-heat effect during multiple layers welding. a) Fine grains structure. b) Intermediate size grains structure. c) Coarse grains structure.   119

Fig. 4.31: Microstructure of weld metals after subcritical PWHT conditions: a) L760, 95% ferrite (F); b) M760, 92% % F; c) H760, 93% F.   121

Fig. 4.32. Microstructure of weld metals: (a) M350. (b) L760, (c) M760, (d) H760. Some ferrite grains are seen in the microstructure.   122

Fig.4.33: Relation between measured ferrite content, resulted hardness and heat input for welds in as-welded condition.   123

Fig. 4.34 : Relation between measured ferrite content, resulted hardness and heat input for HAZ in as-welded condition.   123

Fig 4.35: Relationship between measured ferrite content and hardness for as welded condition, tempered condition and normalized/tempered conditions.   124

Fig. 4.36: SEM (a to c) Microstructure of Normalized/tempered Weld Metals: a) L1050, 94.6% Ferrite (F). b) M1050, 94.4% F. c) H1050, 94.8% F. d) H1050, 94.8% F light micrograph.   125

Fig. 4.37: X-ray diffraction pattern of base metal and weld metals of all heat/treatment conditions.   126

Fig. 4.38: FWHM resulted from X-diffraction analysis conducted on weld metals of different heat input/heat treatment conditions.   126

Fig. 4.39: Crystallite Size resulted from X-diffraction analysis conducted on weld metals of different heat input/heat treatment conditions.   127

Fig. 4.40: Different types of formed δ-ferrite within the weld metals of as-welded conditions: (a-d) L350. (e, f) M350. (h, i) H350.   128

Fig. 4.41: δ-ferrite within the weld metal after subcritical PWHT: a) L760. b) M760. c) H760.   129

Fig. 4.42: δ-ferrite at fusion line and CGHAZ of different heat inputs of the as-welded and subcritical PWHT conditions: Optical micrographs; (a) L760. (b) M350. (C) H350. (d) SEM micrograph, fusion line, H760.   130

Fig.4.43: δ-ferrite within the weld metal after Normalizing and tempering treatment: a) L1050. b) M1050.   c) H1050. — 130

Fig. 4.44: SEM Microstructure of Delta Ferrite in the Weld Metal of L760 and location of EDS analysis conducted (Table 4.10). (a) Delta Ferrite with sub-grains structure. (b) Fine Ferrite grains with excessive precipitates at the boundary of delta ferrite. (c) Sub-grain inside delta ferrite grain with different types of precipitates. — 132

Fig. 4.45: SEM Microstructure of δ-ferrite in the Weld Metal of M760 and location of EDS analysis conducted (Table 4.11). (a) δ-ferrite. (b) Grains with excessive precipitates along the boundary of δ-ferrite. — 133

Fig. 4.46: SEM Microstructure of δ-ferrite in the Weld Metal of H760 and location of EDS analysis conducted (Table 4.13). — 133

Fig.4.47: SEM Microstructure of δ-ferrite in the Weld Metal of M1050 and Locations of conducted EDS analysis. (a) δ-ferrite with fine distributed carbides along the matrix. (b) Ferrite with low precipitation percentage along the δ-ferrite boundary — 135

Fig. 4.48: Different areas at the HAZ of low heat input (L350): (a) CGHAZ. (b) FGHAZ e. (c) ICHAZ. (d) Over tempering zone (OT). (e) Boundary between over tempered (OT) and un-affected base metal (UA). (f) Un-affected Base Metal. — 136

Fig. 4.49: Optical Micrographs of weld metals of normalized/tempered conditions: (a) L1050.   (b) M1050.   (C) H1050. — 138

Fig. 4.50: Optical Micrograph of HAZ after N&T treatment: (a) L1050. (b) M1050. (c) H1050. — 139

Fig. 4.51: SEM showing bainite like grains in the subcritical PWHT condition. (a) HAZ area near base metal of high heat input condition H760. (b) Base metal affected by double tempering at 760 °C. — 140

Nomenclature

ASME	American society of mechanical engineering
BM	Base metal
CGHAZ	Coarse-grained heat-affected zone
Cr_{eq}	Chromium equivalent
EDS	Energy dispersive X-ray
FGHAZ	Fine-grained heat-affected zone
GTAW	Gas metal arc welding
HAZ	Heat-affected zone
HV	Vickers hardness
ICHAZ	Inter-critical heat affected zone
LMP	Larson-Miller parameter
PM	Parent metal
PWHT	Post-weld heat-treated
NIMS	National Institute for Material Science
SEM	Scanning electron microscope
SMAW	Shielded metal arc welding
TEM	Transmission electron microscopy
Tm	Melting Temperature
WM	Weld metal
wt%	Weight percent
δ-ferrite	Delta ferrite

Abstract

This book gives an extensive over view on the material used for the construction of power generation plants, its evolution, types and grades, with comparison for their chemical, mechanical, creep and corrosion resistance. On the other hand, this book focusing also on steel P91 which is known for its excellent high-temperature properties. P91 steel always acts as one of the best material selection for the most critical power boiler components. The achievement of optimum weld metal properties for steel P91 within the course of its extensive applications in power plants has often caused concern. In the present work, modified 9Cr-1Mo steel (P91) were welded using different levels of heat inputs. Three different conditions of heat treatments were employed; as welded, subcritical post-weld heat treatment and normalizing/tempering treatment. Microstructure was evaluated by optical, scanning electron microscopes, X-ray diffraction and Ferrite-scope measurement. Hardness, tensile, impact toughness, stress rupture properties and stress corrosion cracking were evaluated. The best combination of mechanical and corrosion properties with a significant increase of about 90% in time to rupture was obtained at specific heat input and heat treatment condition. Specific heat input and heat treatment combination given the smallest proportion of δ-ferrite in the weld area with great homogeneity of the microstructure were determined. This combination also leads to the absence of the soft over tempered area in the heat affected zone which act as the normal source of failure during services, due to its un-preferred properties.

Chapter 1 : Introduction

Usually, the first chapter of the thesis provides an introduction to the research work. Each chapter may start with an introductory paragraph right after its title to provide some information about its content.

The need to reduce the fuel cost as well as environmental pollution from fossil fuels by significantly decreasing carbon dioxide emissions from power-generation plants has lead to efforts to increase the thermal efficiency of power plants [1]. The increase in the thermal efficiency of fossil fuel fired steam power plant that can be achieved by increasing the steam temperature and pressure; has provided the incentive for the development of heat resistant steels with excellent creep properties as well as superior oxidation and corrosion resistance properties [2].

In the last two decades and to face up these requirements; several new Cr-Mo and 9-12% Chromium steels were developed ranging from P11 (1Cr-0.5Mo) to P122 (12Cr-1Mo) [1&3].

In the USA, a large program was launched by The Electric Power Research Institute (EPRI) in the early 1980s to develop new 9-12% Cr steel and new component designs. Japanese and European companies also jointed the EPRI programs and the 9% Cr steels, P91, P92 and E911, demonstrate what can be achieved through such international efforts [4]. During the last twenty years, three such steels, P91 (9Cr-1Mo-V-Nb), E911 (9Cr-1Mo-1W-V-Nb) and P92 (9Cr-0.5Mo-1.8W-V-Nb), have been developed to commercial production [2].

P91 was originally developed in USA by the Oak Ridge National Laboratory (ORNL) for application in the fast breeder reactor and has now become a well established steel for power station components. ORNL developed this type of steel by increasing the nitrogen content of the basic 9C-1Mo (P9) steel composition and adding small amounts of Vanadium (V) and Niobium (Nb) [1].

There are extensive and reliable design data available, which allow the 10^5 h rupture strength at 600°C to be confidently specified; P91 is covered by the American standards ASTM 213 and 335 [5]. However, the work under ERPI's RP1403 project on super 9Cr (That become P91) led to its being selected as an ASME standardized material in 1984; which clearly stronger than the commonly accepted alloys P11 and P22. By 1997, P91 had been installed in power plants across the world. In 1991, Lower Colorado River Authority replaced P11 secondary superheater outlet header that had shown signs of swelling, with header made of P91 steel. In 1992, Mannesmann supplied P91 headers as replacements for P22 headers at Dayton Power and Light's Stuart plant in Aberdeen, Ohio. In 1993, San Diego Gas and Electric replaced P11 main steam line fittings with P91 in Encina units 4 and 5, when evidence appeared of circumferential weld cracking. Also, in the same year, Consumers Power replaced two P22 secondary superheater outlet headers with a single P91 header at unit 2 of the J.H. Campbell station, subsequently reported in 1997 to be operating successfully with no evidence of cracking. Generally, it was indicated that; initial operating experience was positive with P91 [6].

Following P91, two new grades have been developed nearly simultaneously; P92 which is a Japanese development and steel E911 which has been developed during the European COST (European Cooperation in Science and Technology) activities. The higher creep strength in comparison to P91 was made possible by the addition of tungsten. E911 is alloyed with approximately 1% W. P92 has a higher W-content of 1.7%. At the same time the Mo-content is reduced down to 0.5%, to suppress the formation of delta ferrite. The service temperature of these two grades is in the range of about 585 to 625 °C [7].

In fact, Europeans still aiming for 650 to 700 °C steam, and all over the world numerous research projects are running to gain this aim. In Europe, one important research projects concentrates in the COST Program, where turbine and boiler makers, steel makers, producer of filler metals, research and university institutes and finally operating companies of fossil fuel fired power plants work together to improve material properties. At present time COST 536 program is running. Within this project about 50 partners of more than 15 European countries joined together in five different work packages: alloy design, modeling, turbine group, boiler group and welding group. It is their aim to develop and to qualify creep resistant Ferritic and Martensitic 9-12% Cr materials for the use up to 650 °C steam temperature [8].

The available literatures concentrates on many subjects related to the 9-12%Cr steels material behaviors like:

- Microstructure and carbides formation at elevated temperature.
- Microstructure and carbides stability; and its effect on short and long-term creep strength.
- Creep mechanisms for these types of materials at elevated temperature.
- Welding repair assessment.

While few data are published on the welding characteristics, weld creep behaviors, microstructure, carbides stability, and steam oxidation resistance at elevated temperature. The results of literatures that focused on the weld properties conducted on simulated specimens and a few of them conducted on an actual welds in a low range of heat inputs or without the consideration of the used heat input at all. On the other hand, subcritical PWHT was considered always as the mandatory treatment during the study.

The aim of this work is to give more focus on the welding characteristic comparing to the base material using the most common welding processes (eg. TIG and SMAW). Martensitic Steel P91 (9Cr-1Mo-V-Nb) was selected to use for this aim; as it is the most used steel for the design of different components of the power boilers.

In this study, welded joints were produced in P91 using three different heat inputs, and two different PWHTs (Subcritical PWHT and normalizing/tempering) were carried out for different heat input conditions.

The main subjects of this work is to:

1. Study the combination effect of heat input and heat treatment on the microstructure and mechanical properties of weld metal and HAZ.
2. Study the stability and effect of the resulted microstructure and formed carbides on the creep resistant prosperities through short-term testing for each heat input/treatment condition.
3. Study and compare the resulted microstructure and mechanical properties of normalizing and tempering to that obtained from subcritical post weld heat treatment.
4. Study the relation between stress corrosion cracking resistance and heat input/heat treatment conditions.
5. Compare the results with that of the base material.

This thesis is divided into four chapters. Following the introduction; chapter 2 covers the literature review in which the metallurgy of 9-12% Cr steels, weldability, creep and other properties at high temperature, and steam oxidation resistance are reviewed. Chapter 3 deals with the material and experimental work, and chapter 4 deals with the results and discussion. Finally, summary and conclusions are given.

Chapter 2 : Literature Review

2. 1. Materials for power generation plants-over view:

For most of the materials, mechanical properties are temperature dependent. Strength of materials generally decreases with increasing temperature. Figure 2.1 shows the temperature dependence of tensile strength of some metallic materials. In stress carrying structures creep becomes significant at high temperatures (> 0.4 Tm). Furthermore, rapidly alternating temperatures induce thermal stresses which may cause a premature failure by brittle fracture or thermal fatigue mechanisms **[9]**.

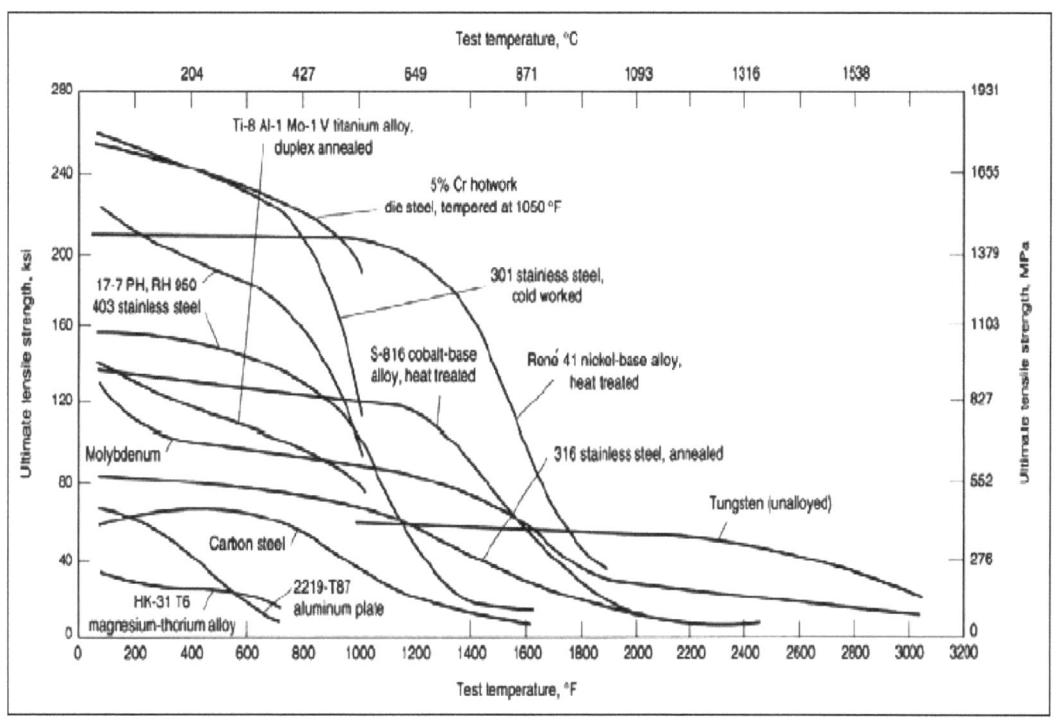

Fig. 2.1: Temperature dependence of tensile strength for selected metallic materials [9].

If thermal expansion of materials is restricted, thermal shocks or temperature changes can generate thermal stresses which may lead to component failure.

Thermal stresses are typically most pronounced at the material surface. Rapid temperature changes at the surface may occur e.g. during shutdowns or when gases of different temperatures are turbulently mixed. Materials resistance to thermal stress-induced failure is raised by high fracture strength and thermal conductivity, low modulus of elasticity and coefficient of thermal expansion, and low heat transfer rate. Unfortunately, materials that typically possess good high temperature stiffness and resistance to environmental degradation (ceramics) are most susceptible to thermal-shock induced failure [10].

In the case of alternating stresses thermal fatigue may occur. Thermal fatigue cracks typically initiate within plastically deformed zone. Therefore, high yield strength improves resistance to thermal fatigue. Good ductility and toughness are also beneficial in this sense. increase of steel strength; consequently, lead to the use of thinner steel thickness, so, the through-wall temperature gradients will be lowered, giving a reduction in the thermal fatigue loading experienced [20]. Figure 2.2 shows the variation in tube wall thickness of six steel grades for service at 593.5°C & 320 bar [6].

At elevated temperatures and corrosive environments damage phenomena such as hot corrosion, oxidation, creep, and thermal fatigue may occur [11]. They are explained in more detail through coming paragraphs within this chapter.

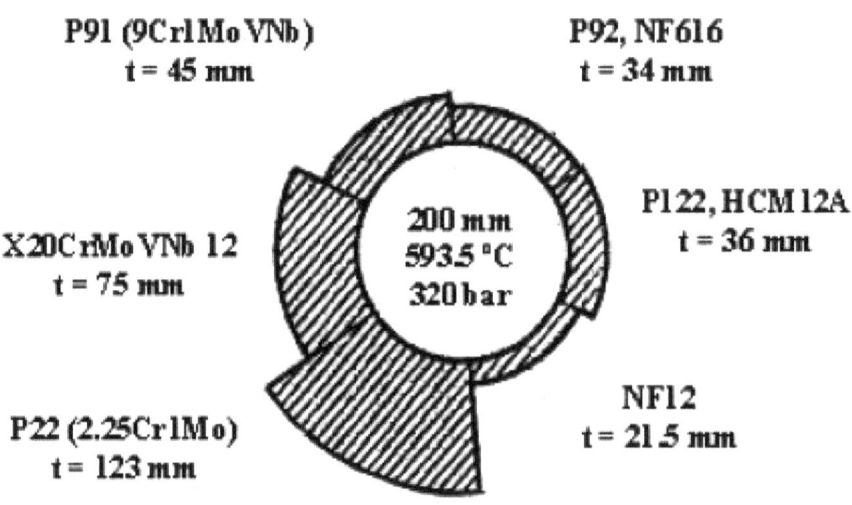

Fig. 2.2: The variation in wall thickness with material grade (with an internal diameter of 200 mm) for service at 593.5 °C & 320 bar.

In high temperature service (345 to 815 °C); and in the design of pressure vessels and piping, engineers and designers are confronted with the problem of selecting materials for a wide range of high-temperature service conditions. The chromium-molybdenum ferretic steels and austenitic stainless steels are generally used for design temperatures above 425°C [12]. In addition to service temperature, corrosion resistance, and fabricability, the following conditions should be considered in high-temperature applications:

- Possible maximum temperature
- Type and size of load
- Expected life of the structure
- Cost

Service experience and laboratory test data have established the normal temperature range of usage shown in table 1 for the materials commonly used for high-temperature service. The allowable stresses in the ASME code, section VIII, division 1, for the low and high ends usage temperature ranges are also given in table 2.1.

Table 2.1: Pressure vessel Steels for high-Temperature Service [12]

ASME NO.	ALLOY AND NOMINAL COMPOSITION	NORMAL TEMPERATURE RANGE OF USAGE		ALLOWABLE STRESS AT LOW END OF USAGE TEMPERATURE		ALLOWABLE STRESS AT HIGH END OF USAGE TEMPERATURE	
		°C	°F	Mpa	Ksi	Mpa	Ksi
SA 204, GRADE C	0.28% C MAX; 0.45-0.60% MO	430-510	800-950	130	18.8	57	8.2
SA 302, GRADE B	0.25% C MAX; 1.15-1.50% MN; 0.45-0.60% MO	430-510	800-950	130	18.8	57	8.2
SA 387, GRADE 12							
CLASS 1	1.0CR-0.5MO	455-565	850-1050	92	13.4	30	43
CLASS 2	1.0CR-0.5MO	345-480	650-900	112	16.3	104	15.1
SA 387, GRADE 112							
CLASS 1	1.25CR-0.5MO	455-565	850-1050	101	14.6	32	4.6
CLASS 2	1.25CR-0.5MO	345-480	650-900	130	18.8	110	15.9
SA 387, GRADE 22							
CLASS 1	2.25CR-1.0MO	455-595	850-1100	99	14.4	40	5.8
CLASS 2	2.25CR-1.0MO	370-480	700-900	119	17.2	109	15.8
SA 387, GRADE 5							
CLASS 1	5.0CR-0.5MO	480-620	900-1150	83	12.1	29	4.2
CLASS 2	5.0CR-0.5MO						
SA 387, GRADE 9	9.0CR-1.0MO	510-595	950-1100	73	10.6	22	3.3
SA 387, GRADE 91	9.0CR-1.0MO + NI, V, NB, N, AL	540-650	1000-1200	99	14.3	28	4.3
SA 240	AUSTENITIC STAINLESS STEELS						
GRADE 304H	18CR-8NI	595-815	1100-1500	61	8.9	9.7	1.4
GRADE 316H	16CR-12NI-2MO	595-815	1100-1500	71	10.3	8.3	1.2
GRADE 321H	18CR-10NI-TI	595-815	1100-1500	48	6.9	2	0.3
GRADE 347H	18CR-10NI-NB	595-815	1100-1500	90	13.0	9	1.3

In fact, the development of power plant technology towards larger units and higher efficiencies is linked to the development of creep resistant ferritic steels. Starting with simple CMn- steels, creep strength has improved successively by introducing new alloying elements and new microstructures. Niobium has been one of the most successful new elements. It is contained in all the latest high strength steels belonging to the group of 9-12%Cr- steels. The use of these steels allows the design of power plants with steam temperatures up to 625°C. As a further step ferritic steels are currently under development for maximum steam temperatures of 650°C, which is believed to be the limit that can be achieved for this group of steels [13].

Table 2.2 gives an overview on creep resistant ferritic steels which are used in power plants for tubing and piping. The list of steels can be subdivided into CMn- steels, Mo-steels, low alloyed CrMo- steels, and 9-12%Cr- steels.

Table 2.2: Chemical compositions of creep resistant ferritic steels for power plants [13].

No.	EN-Designation	Comparable ASTM Grade	Chemical Composition (mass-%)												
			C	Si	Mn	Al	Cu	Cr	Ni	Mo	W	Ti	V	Nb	B	N
1	P 235	A	max. 0.16	max. 0.35	0.40 -0.80	min. 0.020	max. 0.30	max. 0.30	max. 0.30	max. 0.08						
2	P 355		max. 0.22	0.15 -0.35	1.00 -1.50	max. 0.060								0.015 -0.10		
3	16Mo3		0.12 -0.20	0.15 -0.35	0.40 -0.80	max. 0.040				0.25 -0.35						
4	9NiCuMoNb5-6-4		max. 0.17	0.25 -0.50	0.80 -1.20	max. 0.050	0.50 -0.80	max. 0.30	1.00 -1.30	0.25 -0.50				0.015 -0.045		
5	13CrMo4-5	T/P11	0.10 -0.17	0.10 -0.35	0.40 -0.70	max. 0.040		0.70 -1.10		0.45 -0.65						
6	11CrMo9-10	T/P22	0.08 -0.15	0.15 -0.40	0.30 -0.70	max. 0.040		2.00 -2.50		0.90 -1.20						
7	8CrMoNiNb9-10		max. 0.10	0.15 -0.50	0.40 -0.80	max. 0.050		2.00 -2.50	0.30 -0.80	0.90 -1.10				min. 10x%C		
8	7CrMoVTiB10-10	T/P24	0.05 -0.10	0.15 -0.45	0.30 -0.70	max. 0.020		2.20 -2.60		0.90 -1.10		0.05 -0.10	0.20 -0.30		0.0015 -0.007	max. 0.010
9		T/P23	0.04 -0.10	max. 0.50	0.10 -0.60	max. 0.030		1.90 -2.60		0.05 -0.30	1.45 -1.75		0.20 -0.30	0.02 -0.08	0.0005 -0.006	max. 0.030
10	X11CrMo9-1	T/P9	0.08 -0.15	0.25 -1.00	0.30 -0.60	max. 0.040		8.0 -10.0		0.90 -1.00						
11	X20CrMoNiV11-1		0.17 -0.23	0.15 -0.50	max. 1.00	max. 0.040		10.0 -12.5	0.30 -0.80	0.80 -1.20			0.25 -0.35			
12	X10CrMoVNb9-1	T/P91	0.08 -0.12	0.20 -0.50	0.30 -0.60	max. 0.040		8.00 -9.50	max. 0.40	0.85 -1.05			0.18 -0.25	0.06 -0.10		0.030 -0.070
13	X11CrMoWVNb9-1-1	T/P911	0.09 -0.13	0.10 -0.50	0.30 -0.60	max. 0.040		8.50 -9.50	0.10 -0.40	0.90 -1.10	0.90 -1.10		0.18 -0.25	0.06 -0.10	0.0005 -0.005	0.050 -0.090
14		T/P92	0.07 -0.13	max. 0.50	0.30 -0.60	max. 0.040		8.50 -9.50	max. 0.40	0.30 -0.60	1.50 -2.00		0.15 -0.25	0.04 -0.09	0.001 -0.006	0.030 -0.070
15		T/P122	0.07 -0.13	max. 0.50	max. 0.70	max. 0.040	0.30 -1.70	10.0 -12.5	max. 0.50	0.25 -0.60	1.50 -2.50		0.15 -0.30	0.04 -0.10	max. 0.005	0.040 -0.100

2.1. 1. Carbon manganese-steels (CMn- steels) :

Steel grade P 235 can be regarded as typical of a CMn- steel having a ferrite perlite microstructure. The carbon and manganese contents are the major factors influencing the strength properties. Figure 2.3 shows a plot of the minimum 0.2%- proof strength values together with the average 10^5 h- creep rupture strength as function of temperature. The European design codes are based on minimum proof strength values at low and creep rupture strength values at high temperatures. Both regimes are separated by the intersection of the proof strength with the creep rupture strength curve. In the case of P235 the intersection point is around 420°C. Above this temperature design becomes time dependent, because the life time of a component is limited by the creep process.

An interesting modification of P 235 is the niobium bearing grade P 355. The proof strength values could be raised considerably as a result of niobium addition due to grain refinement. However, the increase of creep rupture strength is rather small. This increase can be attributed mainly to the increase of manganese, which is a solution hardening element. Since proof and creep rupture strength have not increased by equal amounts, the intersection point between the creep and proof strength regime is shifted to 400°C. The advantage of P 355 clearly lies in the application below this temperature.

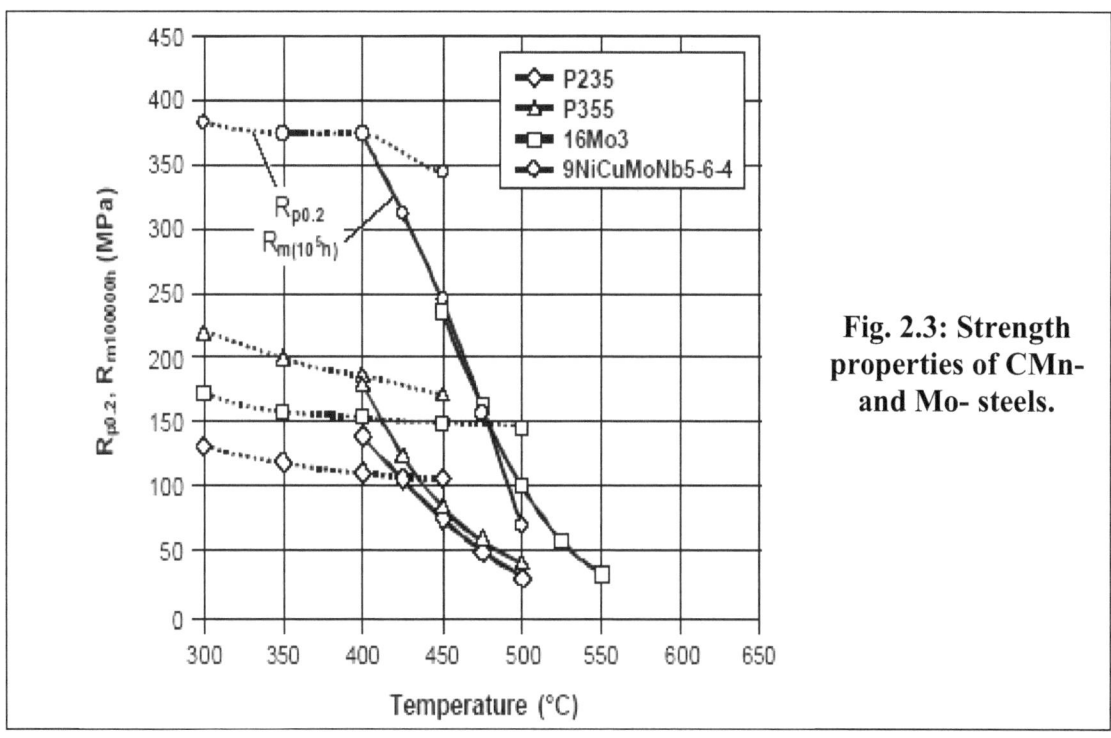

Fig. 2.3: Strength properties of CMn- and Mo- steels.

2. 1. 2. Mo- Steels:

A similar effect can also be seen for the Mo- steels, which are also represented in Figure 2.2. These steels are basically of the same type with about 0.3% molybdenum, which is a strong solution hardening element. The solution hardening effect is the main cause for the increase of creep rupture strength, which is similar for both steels. Grade 9NiCuMoNb5-6-4, also well known as WB 36, shows a dramatic increase of proof strength, which again is partly caused by the grain-refining effect of niobium. In addition hardening by copper precipitation increases the proof strength.

2. 1. 3. Cr Mo- Steels:

It has been found that the strengthening effect of molybdenum cannot fully be used, since creep ductility strongly decreases with increasing molybdenum content. Another limitation in the application of Mo- steels is the observed decomposition of iron carbides above 500°C (graphitization). A solution to both problems was the use of chromium as an alloying element in combination with molybdenum. In fact CrMo- steels were the first ones that allowed steam temperatures in power stations above 500°C. The classical CrMo- steels are 13CrMo4-5 (T/P11) and 11CrMo9-10 (T/P22). Their creep rupture strengths are distinctly higher than the simple Mo- steels (Figure 2.4), which is mainly a result of higher Mo- content. CrMo steels form chromium carbides which are stable above 500°C. Therefore graphitization is no longer a problem. Chromium also promotes the use at higher temperatures due to its positive influence on oxidation resistance. The steels 7CrMoVTiB10-10 (T/P24) and T/P23, also represented in Figure 2.4, reveal extremely high strength properties. These are newly developed steels on the basis of T/P22. Having a similar microstructure as T/P22, their strengths properties have been raised considerably by additional alloying with titanium, vanadium and boron in the case of T/P24 as well as tungsten, vanadium, niobium and boron in T/P23.

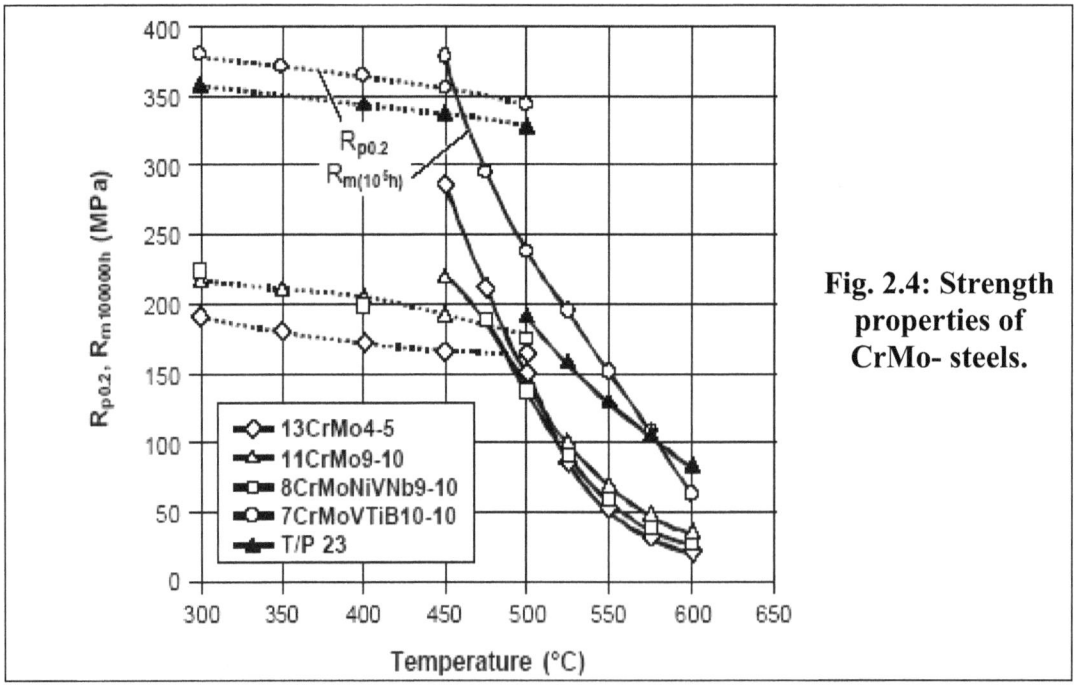

Fig. 2.4: Strength properties of CrMo- steels.

Another interesting steel of this group is 8CrMoNiNb9-10. This is also a niobium bearing steel which reflects another facet of its use in creep resistant ferritic steels. The steel has been developed for nuclear applications in liquid sodium cooled fast breeder reactors.

A problem has arisen from the use of low chromium ferritic steels, like T/P22, in combination with high chromium austenitic steels. The higher carbon affinity of the high chromium austenitic steel caused a massive diffusion of carbon from the ferritic steel via the liquid sodium into the austenitic steel. As a result a strong decrease of strength occurred in the ferritic steel and an embrittlement in the austenitic steel. The problem was solved by alloying T/P22 with niobium leading to the formation of niobium carbides instead of chromium carbides. The stronger affinity of carbon to niobium prevents a carbon depletion of the ferritic steel and consequently a decrease in strength.

2. 1. 4. 9-12% Cr - Steels:

The increase of chromium in CrMo- steels above 7% leads to a group of steels which have a martensitic microstructure as common feature. This microstructure introduces a new element of structural hardening. It is characterized by a high dislocation density and a fine martensite lath structure which is stabilized by $M_{23}C_6$ precipitates. Thus structural hardening is responsible for the large increase in strength of X11CrMo9-1, as compared to 11CrMo9-10 (Figure 2.5).

Further improvements of especially the creep strength have been achieved by alloying with vanadium, niobium, tungsten and boron. The introduction of X20CrMoNiV11-1 at the beginning of the sixties has been a major step to increase power plant efficiency [14]. Its creep rupture strength at a temperature of 540°C is nearly twice that of the low-alloy ferritic steels available at that time (e.g. 10CrMo9-10 with a 10^5 h- creep rupture strength of 78 MPa compared to 147 MPa for X20CrMoNiV11-1). Transformation behavior and microstructure is comparable to X11CrMo9-1. The higher creep rupture strength of X20CrMoNiV11-1 is mainly caused by the higher amount of $M_{23}C_6$ carbides as caused by the higher carbon content.

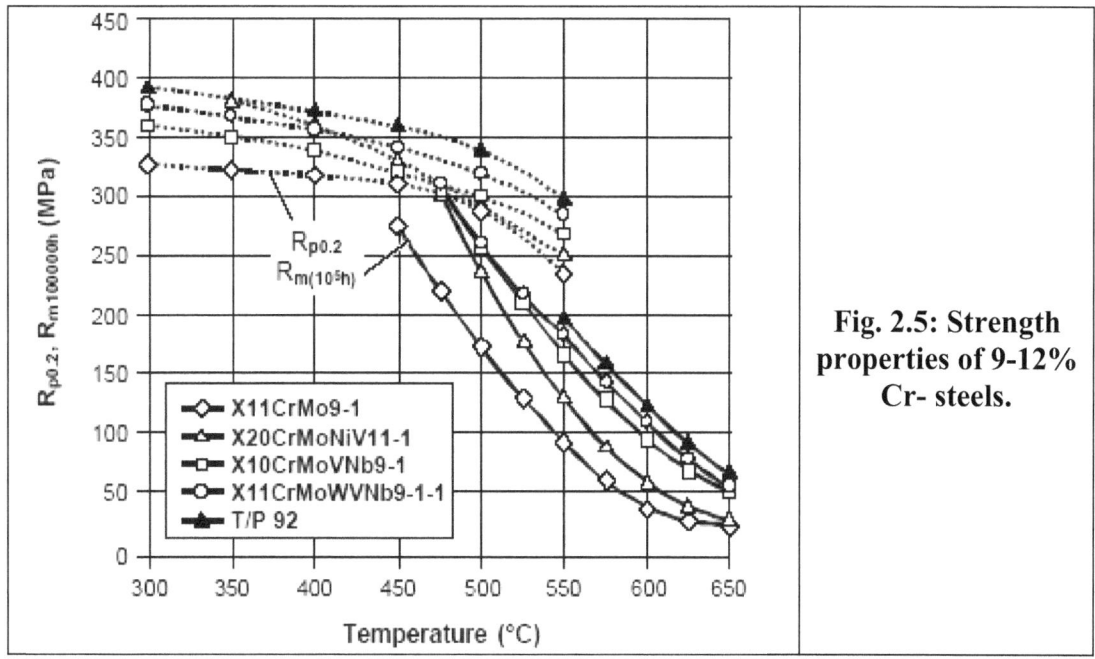

Fig. 2.5: Strength properties of 9-12% Cr- steels.

After a period of standstill, the material development was reactivated by work carried out in the USA and Japan in the mid seventies [15]. The prototype of the new steels from this development work is the modified 9% Cr steel T/P91 (EN designation: X10CrMoVNb91) invented in the USA [16&17]. Meanwhile this steel is well known and applied in power plants all over the world. It is used in new plants as well as in refurbishment work of high pressure/high temperature piping systems. Although the carbon content is lower, the creep rupture strength of T/P91 is distinctly higher than that of X20CrMoNiV11-1. This has been achieved by alloying with vanadium and niobium. T/P91 uses the precipitation of finely dispersed Nb/V carbonitrides of type MX as additional strengthening effect. It was essential to balance the composition, because an optimum dispersion and particle size of MX can only be achieved by an optimized Nb/V- ratio and nitrogen content. Subsequently new steel grades have been developed on the basis of T/P91, like X11CrMoWVNb9-1-1 (T/P911), T/P92 and T/P122 [15]. These steel grades represent the current state of development for creep resistant ferritic steels. Figure 2.6 showing the evolution of these types of ferritic steels in power boiler.

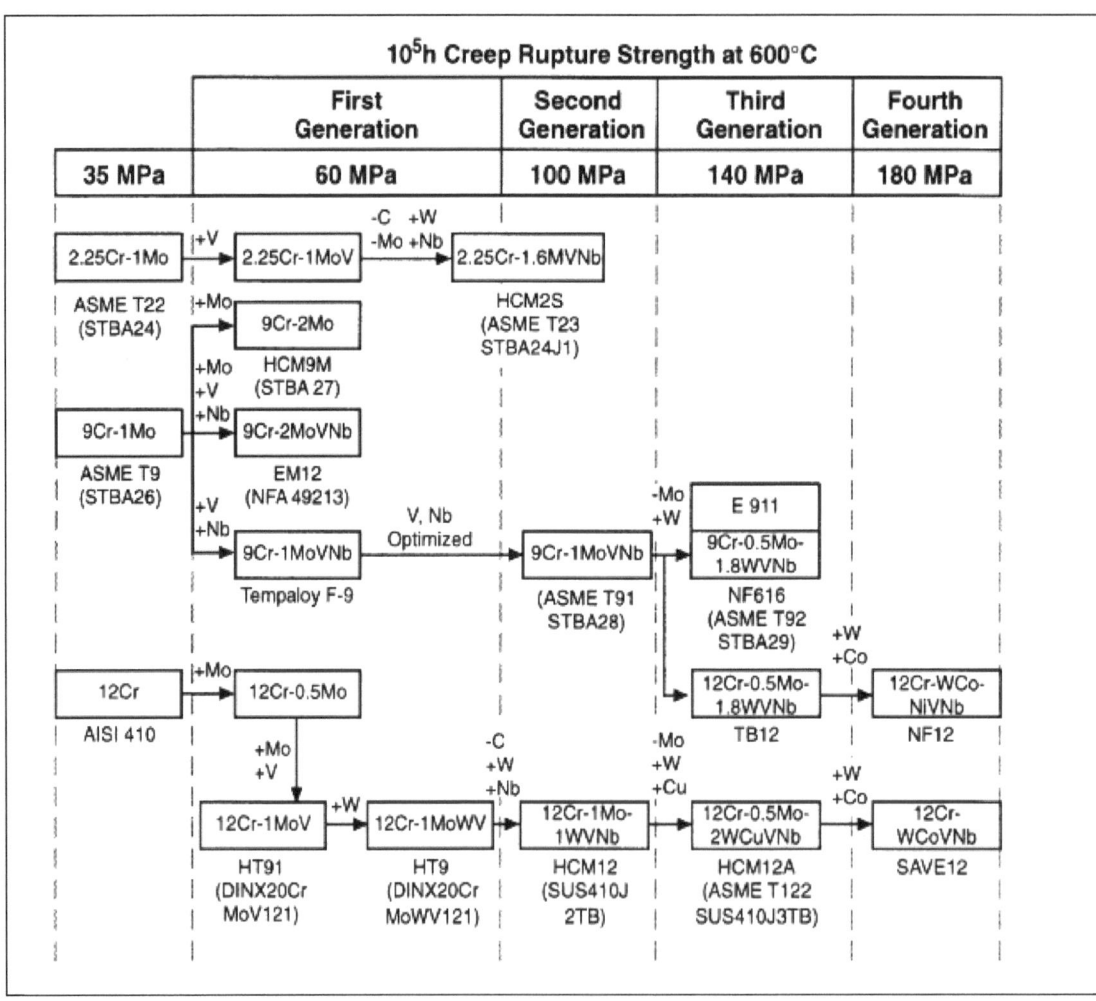

Fig. 2.6: Evolution of ferritic steels for boilers [19].

The properties and suitability of 9-12% Cr-Steels for high temperature application will be discussed in more details through coming titles.

2. 1. 5. High performance creep resistant steels for 21ˢᵗ century power plants:

Research and developments of heat resistant steels and alloys for high-efficient power plants at 650 °C and above are being now promoted in Europe, USA and Japan. Great advancement has been achieved in the analysis of specific microstructure instability causing a loss of creep strength, the prediction of the onset time of the creep strength loss and the theoretical modeling of precipitation sequences in power plant steels. A group of highly-creep resistant martensitic 9Cr steels with higher creep rupture strength than existing high strength steels such as T91 and P92 have been proposed at NIMS (National Institute for Material Science) project in Japan for application to 650 °C USC plants. The formation of thin scale of Cr-rich oxides is achieved on the surface of 9Cr steel by the combination of Si addition and pre-oxidation treatment in argon gas at 700°C. This significantly improves oxidation resistance of 9Cr steel in steam at 650°C. The addition of boron suppresses the Type IV fracture in the heat affected zone at low stresses significantly improves the creep rupture strength of welded joints. Approximately no difference in microstructure of the heat affected zone and base metal produces no mechanical constrain effect in heat affected zone [18]. The future prospect of Austenitic stainless steels and Ni- base alloys for 700°C USC plants is also described; but, the issues of long-term creep properties and manufacturing methods had to be addressed [4].

2. 1. 6. Austenitic stainless steels:

Ni alloying reduces the oxidation rate of Cr steels and transforms the BCC (ferrite) crystal structure to FCC (austenite) structure. Cr-Ni steels have relatively low cost, good mechanical properties, and moderate oxidation resistance. Therefore, they are commonly used in low performance high-temperature applications. Austenitic stainless steels exhibit higher strength and creep resistance at elevated temperatures than non-strengthened ferritic stainless steels, because the crystal structure of austenite is more stable at high temperatures. Also, lattice resistance to dislocation movement is not so temperature sensitive in austenite and diffusion rate is lower than in ferrite [25].

Austenitic steels are candidates primarily in the finishing stages of superheater/reheater tubing, where, oxidation resistance and fireside corrosion become important in addition to creep strength. From a creep strength point of view, T91 is limited to 565°C steam (metal 593°C) and NF616, HCM12A and E911 are limited to 593°C steam (metal 620°C). Even the strongest ferritic steel today is limited to 593°C (1150°F) (metal temperature) from an oxidation point of view. At temperatures above these, austenitic steels are required. Hence there has been considerable development with respect to austenitic stainless steels.

In actual practice in the U.S. SS304M and SS347 are widely used instead of T-91 in superheater applications, mainly because they are easier to weld, while the cost difference is relatively small. Table 2.3 lists the compositions of various stainless steels for Superheater/reheater tube applications. The steels fall into four categories: 15Cr, 18Cr, 20-25Cr and higher Cr stainless steels.

Table 2.3: Nominal Chemical Compositions of Austenitic Steels for Boiler (wt%)[21]

Steels		Specification		Chemical Composition (mass %)											
		ASME	JIS	C	Si	Mn	Ni	Cr	Mo	W	V	Nb	Ti	B	Others
18%Cr–8%Ni	18Cr8Ni	TP304H	SUS304HTB	0.08	0.6	1.6	8.0	18.0	—	—	—	—	—	—	—
	18Cr9NiCuNbN	TP304CuCbN	SUS304J1HTB	0.10	0.2	0.8	9.0	18.0	—	—	—	0.40	—	—	3.0Cu, 0.10N
	18Cr10NiTi	TP321H	SUS321HTB	0.08	0.6	1.6	10.0	18.0	—	—	—	—	0.5	—	—
	18Cr10NiNbTi	—	SUS321J1HTB	0.12	0.6	1.6	10.0	18.0	—	—	—	0.10	0.08	—	—
	16Cr12NiMo	TP316H	SUS316HTB	0.08	0.6	1.6	12.0	16.0	2.5	—	—	—	—	—	—
	18Cr10NiNb	TP347H	SUSTP347HTB	0.08	0.6	1.6	10.0	18.0	—	—	—	0.8	—	—	—
	18Cr10NiNb (FG)	TP347HFG	—	0.08	0.6	1.6	10.0	18.0	—	—	—	0.8	—	—	—
15%Cr–15%Ni	17Cr14NiCuMoNbTi	—	—	0.12	0.5	0.7	14.0	16.0	2.0	—	—	0.4	0.3	0.006	3.0Cu
	15Cr10Ni6MnVNbTi	—	—	0.12	0.5	6.0	10.0	15.0	1.0	—	0.2	1.0	0.06	—	—
20 to 25%Cr	25Cr20Ni	TP310	SUS310TB	0.08	0.6	1.6	20.0	25.0	—	—	—	—	—	—	—
	25Cr20NiNbN	TP310CbN	SUS310J1TB	0.06	0.4	1.2	20.0	25.0	—	—	—	0.45	—	—	0.2N
	21Cr32NiTiAℓ	Alloy 800H	NCF800HTB	0.08	0.5	1.2	32.0	21.0	—	—	—	—	0.5	—	0.4Aℓ
	22Cr15NiNbN	—	SUS309J4HTB	0.05	0.4	1.5	15.0	22.0	—	—	—	0.7	—	0.002	0.15N
	20Cr25NiMoNbTi	—	SUS310J2TB	0.15	0.5	1.0	25.0	20.0	1.5	—	—	0.2	0.1	—	—
	22.5Cr18.5NiWCuNbN	—	SUS310J3TB	0.10	0.1	1.0	18.0	23.0	—	1.5	—	0.45	—	—	3.0Cu, 0.2N
High Cr –High Ni	30Cr50NiMoTiZr	—	—	0.06	0.3	0.2	50.0	30.0	2.0	—	—	—	0.2	—	0.03Zr
	23Cr43NiWNbTi	—	—	0.08	0.4	1.2	43.0	23.0	—	6.0	—	0.18	0.08	0.003	—

FG : Fine Grained

The various stages in the evolution of these steels have consisted of initially adding Ti and Nb to stabilize the steels from a corrosion point of view, then reducing the Ti and Nb content (under-stabilizing) to promote creep strength rather than corrosion, followed by Cu additions for increased precipitation strengthening by fine precipitation of a Cu rich phase [19].

Under-stabilizing is one of the techniques for improving the creep strength of 18%Cr–8%Ni steels. This method enhances creep strength through improvement of precipitation morphology by fixing C in alloys and decreasing carbide forming elements such as Ti and Nb, which hinder Cr carbide formation, to the point where their contents are insufficient for the C fixation. Figure 2.7 shows this, and the peak point of the creep rupture strength against the ratio of (Ti1 0.5Nb)/C is at a position far apart from the peak point of the conventional Type 321 or Type 347 steels, showing that reducing additions of Ti and Nb relative to the C content can be useful [29].

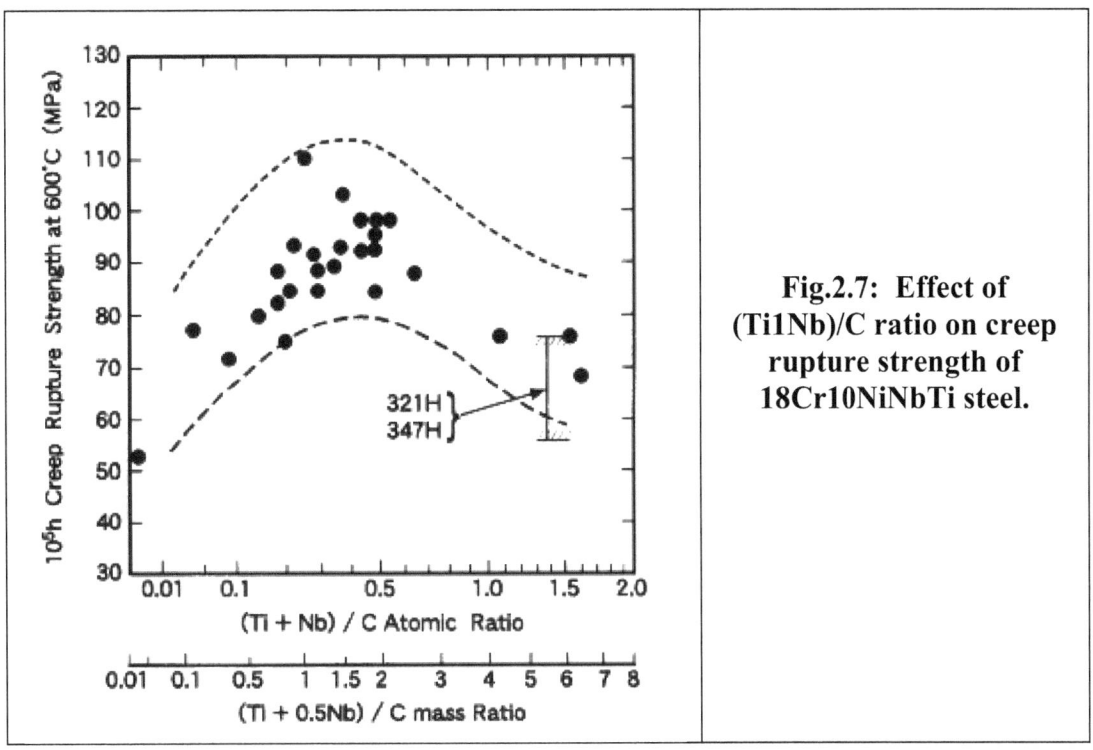

Fig.2.7: Effect of (Ti1Nb)/C ratio on creep rupture strength of 18Cr10NiNbTi steel.

Figure 2.8 shows the effect of the Cu additions on the creep rupture strengths of 18Cr9NiNbN steels. Although the Cu addition does not show a major change up to about 2%, a substantial enhancement in creep strength by means of Cu addition of about 3% or more can be observed. However, because the strength tends to be saturated, and decline in creep rupture ductility can occur when the Cu addition exceeds 3%, the addition of Cu at 3% should be suitable [30].

Further trends have included austenite stabilization using 0.2% nitrogen and W addition for solid solution strengthening [19&41]. This development sequence is illustrated in Figure 2.9.

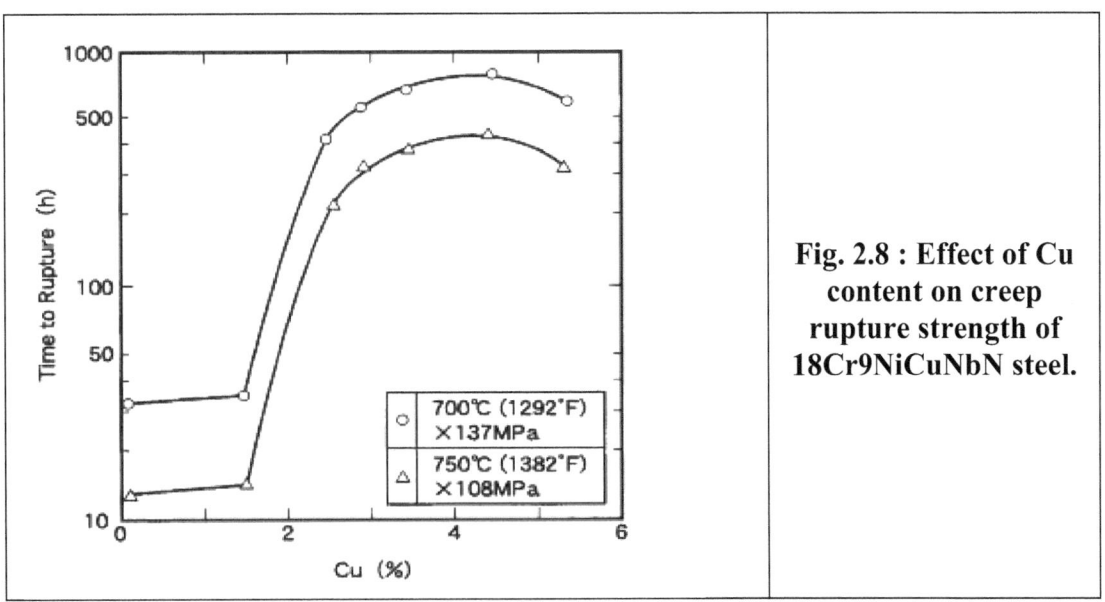

Fig. 2.8 : Effect of Cu content on creep rupture strength of 18Cr9NiCuNbN steel.

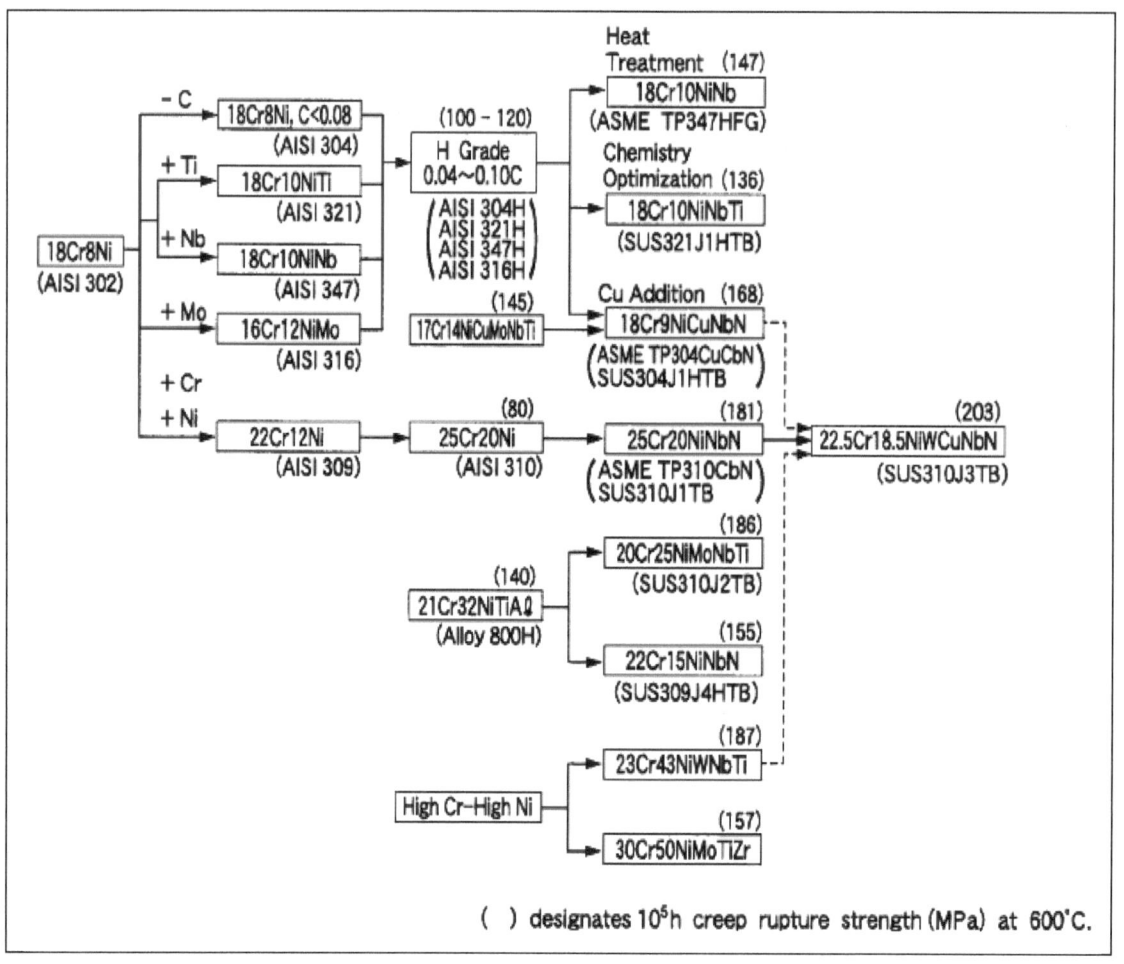

Fig. 2.9 : Evolution of Austenitic steels for boilers [21]

2. 2. Alloy design for 9-12% cr steels :

Much research has been performed since 1960 on the effects of alloying elements on the creep strength of 9–12% Cr steels. Alloying elements for the 9–12%Cr steels are easy to understand if they are grouped in terms of their properties and effects into:
1) Cr; 2) Mo, W, and Re; 3) V, Nb, Ti, and Ta; 4) C and N; 5) B; 6) Si and Mn; and 7) Ni, Cu, and Co.

2. 2. 1. Chromium (Cr) :

Cr is the basic alloying element for heat resistant steels, and increased Cr content improves oxidation and corrosion resistance. Although Cr percent does not exhibit a marked effect on creep strength, high strength is more likely to be obtained near Cr percentages of 2% and 9 through 12% in ferritic steels, and strength declines at compositions between the two coverage's. The reason for this remains unknown [21]. Cr is a ferrite stabilizer and carbides formers; forms hard (often complex) carbides, increasing steel hardness and strength [22]. In practice, Cr is the alloying element most commonly used as a carbide stabilizer [23].

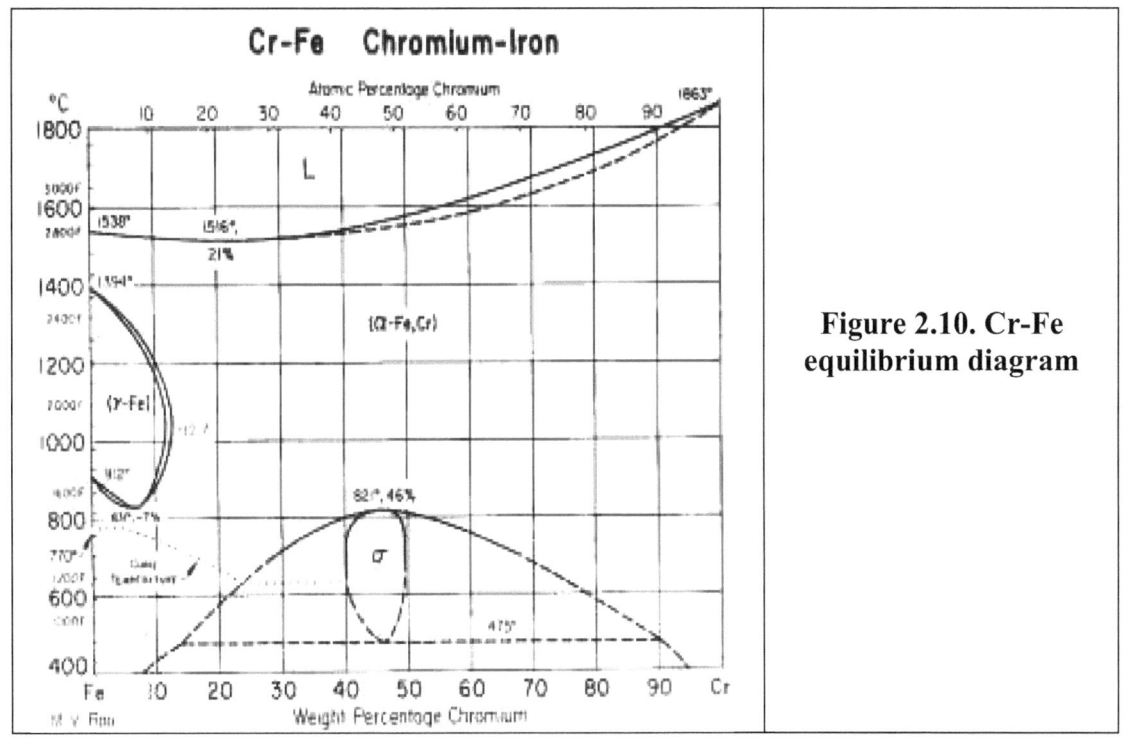

Figure 2.10. Cr-Fe equilibrium diagram

2. 2. 2. Molybdenum, tungsten and rhenium (Mo, W & Re):

Mo, W and Re are all elements useful to solution strengthening. Mo and W have long been used for heat resistant steels. Mo increases hardenability and strength particularly at high temperatures and under dynamic conditions. On the other hand, W increases hardness particularly at elevated temperatures due to stable carbides, refines grain size [22]. Also, these elements further enhance the creep strength of heat resistant steels when added in greater quantities. If their additions exceed a certain limit, however, δ-ferrite precipitates and reduces the strength, and precipitation of the Laves phase decreases toughness. Furthermore, the effect of W on creep strength is approximately half that of Mo, and, as described later, the combined addition of Mo and W can be effective for strength improvement. Re is reported to raise creep strength if added in amount of around 0.5%, and this effect is similar to the actions of Mo and W [21&24].

2. 2. 3. Vanadium, niobium and titanum (V, Nb & Ti):

V, Nb, Ti and Ta all combine with C and/or N to produce carbides, nitrides or carbonitrides, which finely and coherently precipitate on the ferritic matrix to exhibit a marked effect of precipitation strengthening. Among these, V and Nb are found to exhibit particularly optimal contents, about 0.2% and 0.05% respectively, and, as described later, the effect of their combined addition can be great. This suggests that the formations of precipitates composed by V and Nb are associated with each other [21].

2. 2. 4. Carbon and nitrogen (C & N):

Because C and N are austenite formers, they are useful in inhibiting δ-ferrite. Also, their contents relate to the precipitation and coarsening of Cr carbides and nitrides. For C particularly, if addition exceeds 0.1%, the creep strength often declines, and it is believed that there should be an optimal addition according to the types and contents, *etc.* of carbide-forming elements. N is believed to be an element essential for raising creep strength in 9%Cr steels. Additions of N are often at about 0.05%, and it is believed that there should be an optimal content relative to other nitride- forming elements such as B [21]. In fact All carbide formers are also nitride formers.

Nitrogen may be introduced into the surface of the steel by nitriding. By measuring the hardness of various nitrided alloy steels it is possible to investigate the tendency of the different alloying elements to form hard nitrides or to increase the hardness of the steel by a mechanism known as precipitation hardening. The results obtained by such investigations are shown in Figure 4, from which it can be seen that very high hardness result from alloying a steel with Al or Ti in amounts of about 1,5% [23].

On nitriding the base material in Figure 2.11, hardness of about 400 HV is obtained and according to the diagram the hardness is unchanged if the steel is alloyed with Ni since this element is not a nitride former and hence does not contribute to any hardness increase.

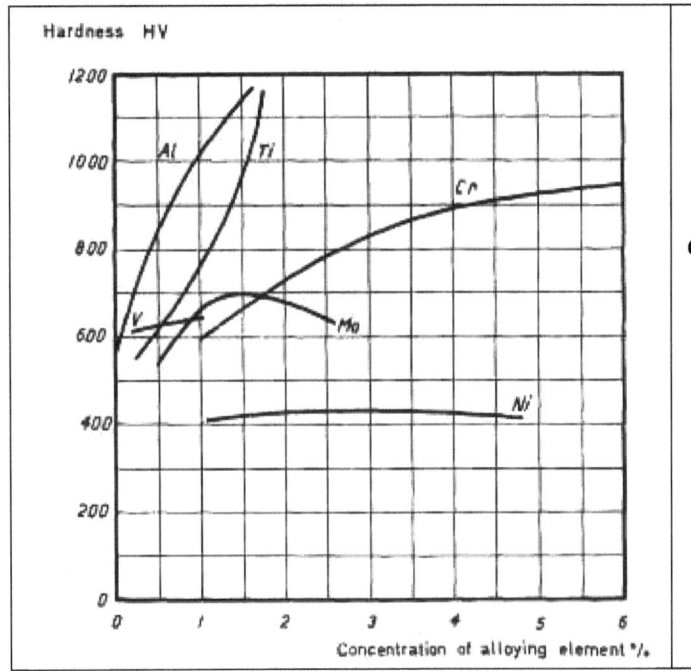

Fig. 2.11 : Effect of alloying element additions on hardness after nitriding[23].

Base composition:
0,25% C, 0,30% Si, 0,70% Mn

2. 2. 5. Boron (B):

B improves hardenability and enhances grain boundary strength, and can greatly improve creep strength. Furthermore, a recent publication indicates that it exhibits the effect of stabilizing carbides by penetrating into $M_{23}C_6$ [21]. This is because the presence of B in the steel could stabilize $M_{23}C_6$ carbide to form $M_{23}(C, B)_6$, resulting in the lowering of its coalescence rate and this improve the long-term creep strength. However, the B addition increases remarkable even a short-term creep strength of steels, which can't be explained by the stabilization of $M_{23}C_6$ carbides alone. *Azuma et al*, have reported that; the B addition changes special distribution of XM carbonitrides in the steels [24].

2. 2. 6. Manganese and silicon (Mn & Si):

With respect to Si and Mn, Si is a ferrite former, whereas Mn is an austenite former. These actions are viewed as being contradictory to each other, and reduction of the contents of both of these elements can improve creep strength. Also, Si works to decrease toughness by promoting the Laves phase, whereas Mn, though useful for toughness improvement, can impair the high temperature stability of the ferrite structure by decreasing the A_1 transformation temperature in the same manner as Ni [21].

2. 2. 7. Nickel, copper and cobalt (Ni, Cu & Co):

Ni, Cu and Co are all austenite formers, and if added as alloy elements, they inhibit the formation of δ-ferrite by decreasing the Cr equivalent, but they simultaneously decrease the A_1 transformation temperature. However, level of this decrease varies among these elements, and the decline seen with additions of Cu and Co is not greater than that found with the addition of Ni. Therefore, if Cu and/or Co are added, the effect of the inhibition of δ-ferrite formation can be expected, making high-temperature tempering possible [21].

2. 2. 8. Combination effect of alloying elements in 9-12% Cr steels:

Of the alloy elements in boiler steels as discussed above, the combination effects of Mo *versus* W and V versus Nb are of interest. As shown in figure 2.12, in the combination of Mo and W, increasing the W ratio while retaining the Mo equivalent (Mo10.5W) at 1.5% is most effective for creep strengthening. Optimal contents of V and Nb may change somewhat according to temperature, and the combination of 0.25% and 0.05%, respectively, as already noted, is optimized for maximum creep strength [26].

Furthermore, interesting findings on rolls of the alloying elements W, Mo and Co to 12%Cr steel for turbine rotors have been reported. Figure 2.13 shows the effects of additions of W1Mo and Co on the creep rupture strength of 12%Cr steel for turbine rotor. The W1Mo content has been changed by increasing W and decreasing Mo on the basis of 1%W, 11%Mo in 12%Cr steel. Time to creep rupture is found to be longest with a combination of 1.8% W and 0.7% Mo, with no decrease in toughness. With regard to Co content, it is found that the maximum time to creep rupture is obtained with a content of 3%, without any major effect on toughness [27].

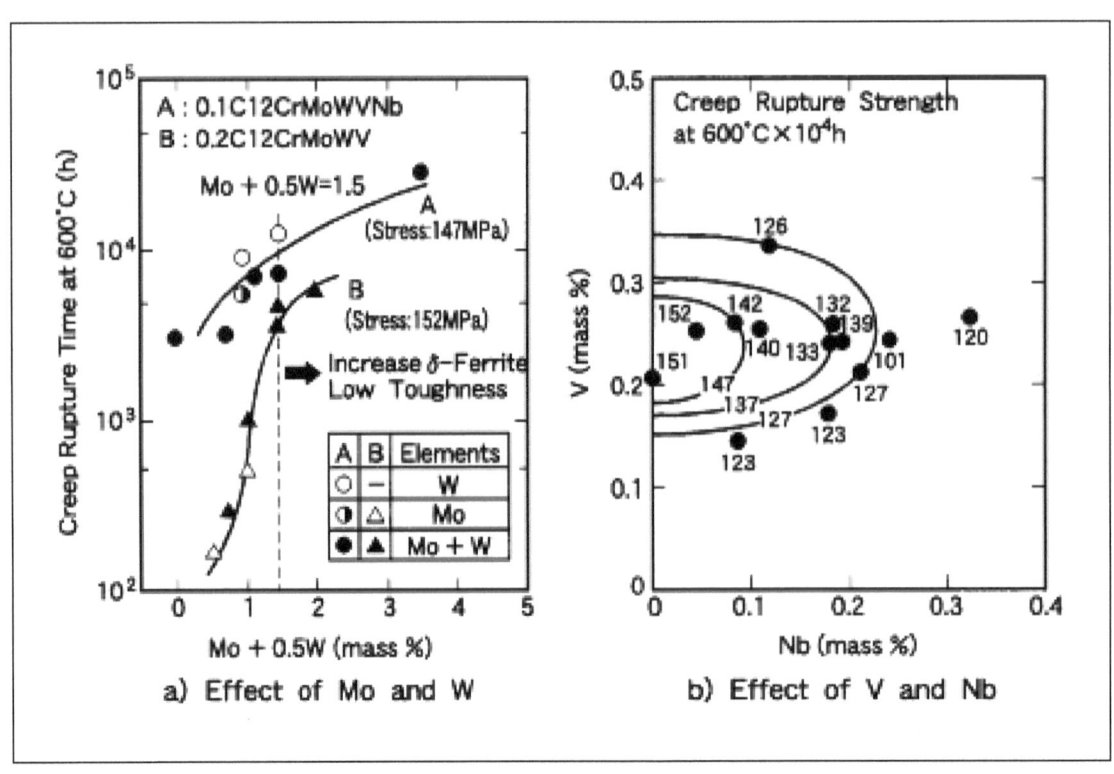

Fig. 2.12: Effect of Mo+W and V+Nb on creep rupture strength of 12%Cr steels [26].

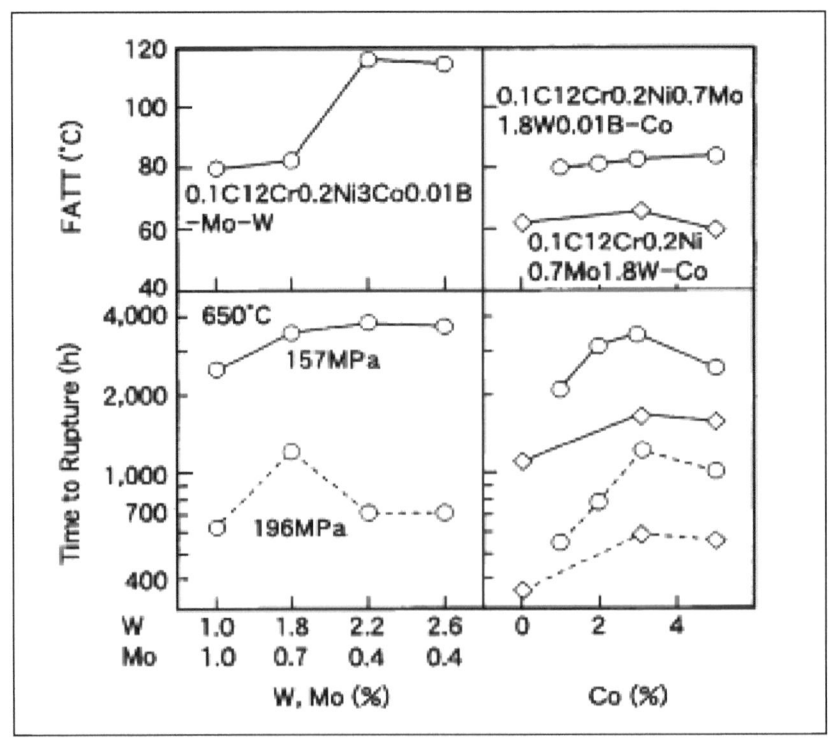

Fig. 2.13: Effect of Mo and W contents on 10^5 h creep rupture strength at 650 °C and Charpy absorbed energy at 20°C of 12%Cr turbine steels (0.13C0.05Si0.5Mn11.2Cr-0.8Ni 0.2V0.05Nb 0.05N–Mo–W).

Figure 2.14 shows the effects of Mo and W contents on 100 000-h creep rupture strength at 650°C and the Charpy impact of 12%Cr steels for turbines. It has been empirically shown thus far that creep strength peaks when the Mo equivalent (Mo10.5W) is set at 1.5 %. It is also known that increasing W content induces increased creep strength and reduced ductility and toughness when the chemical composition is determined along the line connecting 3% W and 1.5% Mo, *i.e.*, such that the Mo equivalent is 1.5% [28].

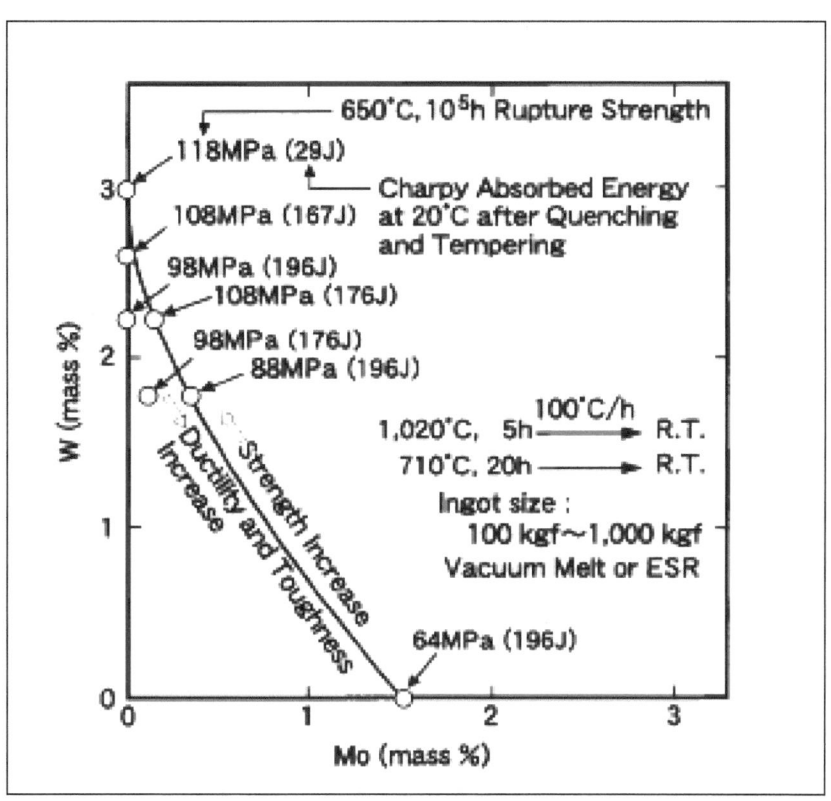

Fig. 2.14: Effect of W1Mo and Co contents on creep rupture strength of 12%Cr turbine steels.

However, Figure 2.15 & 2.16 summarized the concept of alloy design for heat resistant steels to improve creep strength through the modification of existing steels [21].

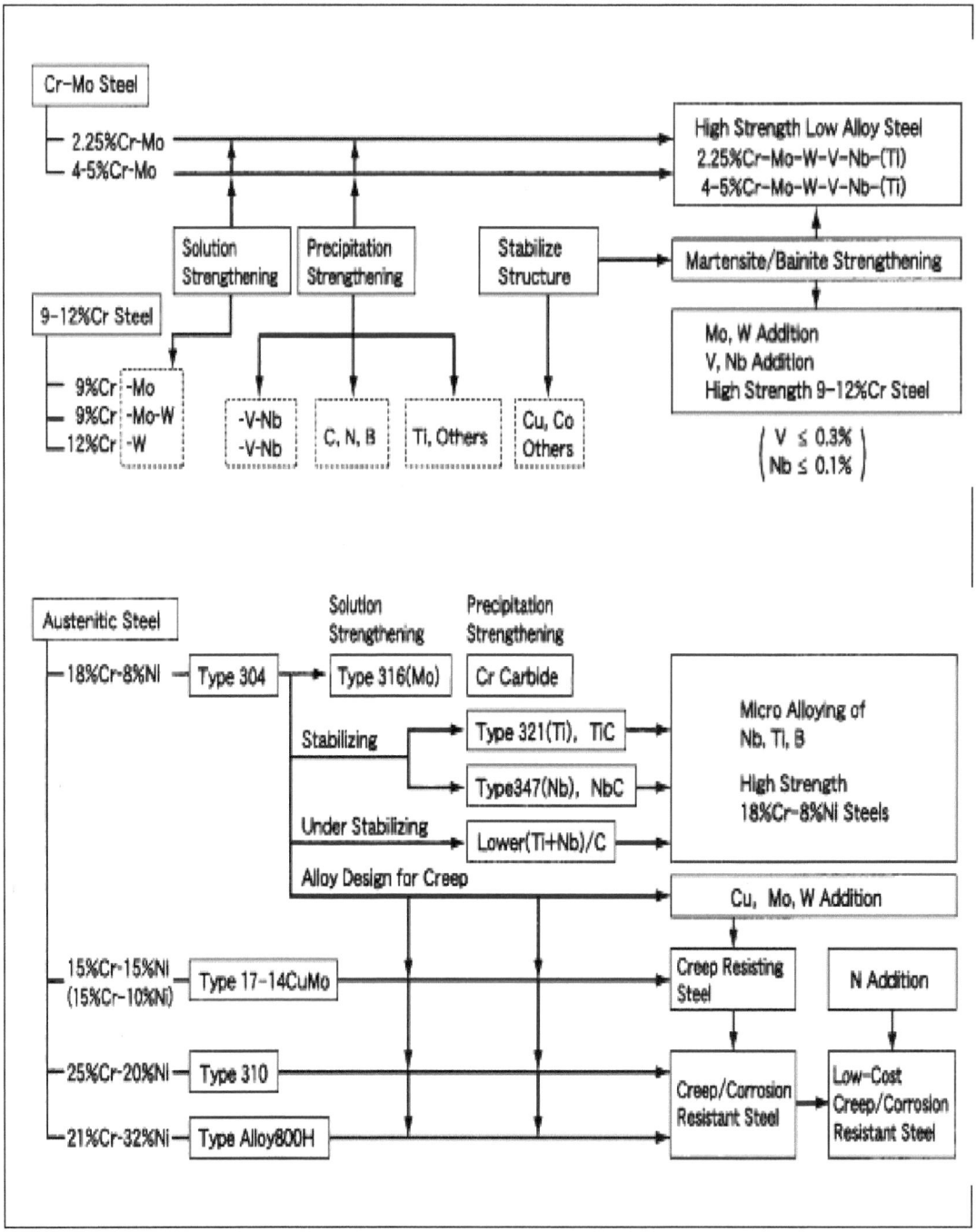

Fig. 2.15: General concept of alloy design for heat resistant steels.

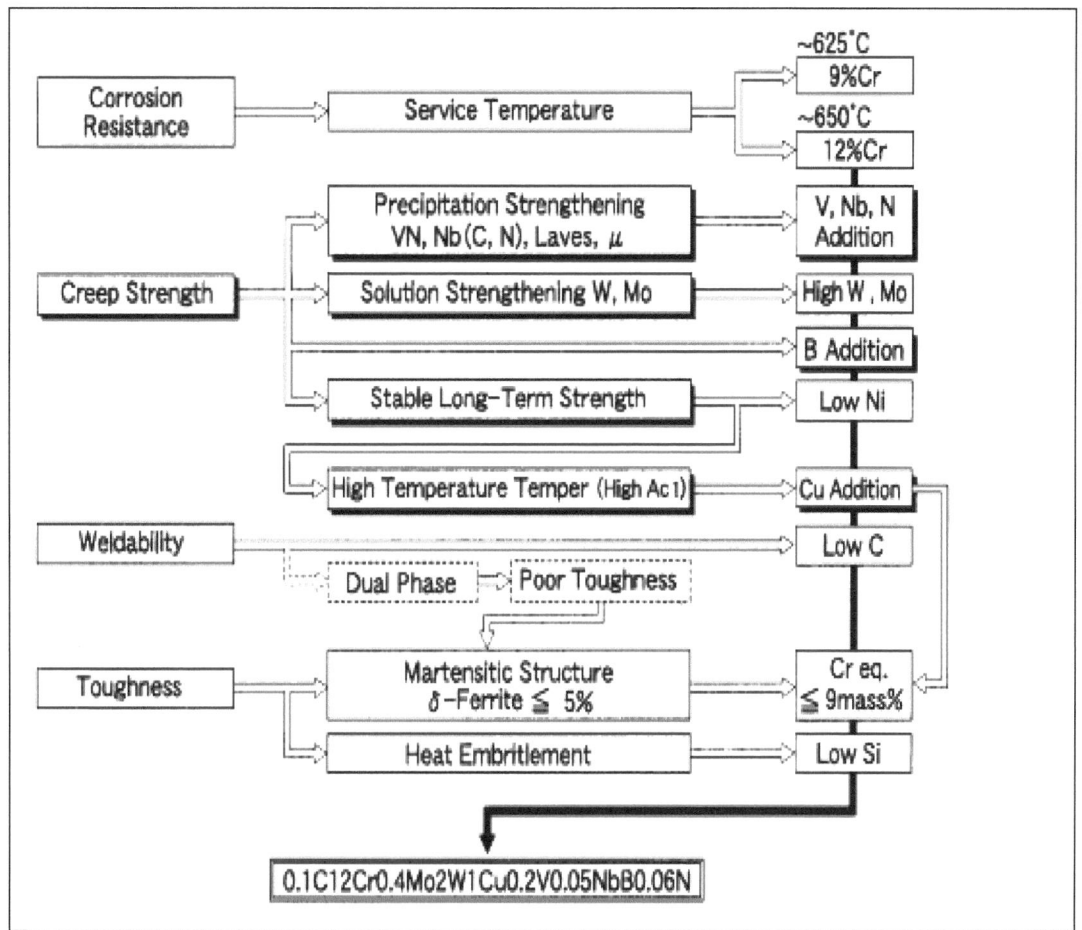

Fig. 2.16: Alloy design of 12Cr0.4Mo2WCuVNb steel.

2. 3. Microstructure of 9-12% cr creep resistant steels:

The microstructure of the 9-10%Cr steels consists of tempered martensite with a high dislocation density and finely distributed carbides, nitrides or carbonitrides. For the application of these materials in steam power plants, the microstructural stability during service is of great importance. The following structural features are expected to exert an influence on the creep rupture properties of the steels [2] :

- The dislocation density within the martensite laths;
- polygonization conditions of the sub-grains;
- fine, uniformly dispersed carbides and carbonitrides within the structure;
- solid solution strengthening of the matrix by elements such as chromium, molybdenum and tungsten;
- formation of inter-metallic phases, such as Laves phase.

Microstructures of 9–12%Cr steels currently being developed or already commercially available consist of a single phase of tempered martensite, with some exceptions. High density dislocations exist in this structure, and the dislocation density is principally influenced by the tempering temperature. It becomes high when the tempering temperature is low, as in the case of turbine rotor steels. Figure 2.17 shows a representative microstructure observed through optical microscopy and transmission electron microscopy (TEM) of a typical 9–12%Cr heat resistant steel. The tempered martensite is composed of numerous laths, and Cr carbides such as $M_{23}C_6$ precipitate along the lath boundaries and along the prior-austenite grain boundaries. Fine MX carbonitrides of (V, Nb)(C, N) coherently precipitate on the ferrite matrix in laths, and dislocation networks are formed along the lath boundaries or the sub-grain boundaries. It is considered that the creep strength of 9–12%Cr steels is closely associated with the stabilization of MX carbonitride and the dislocation structures, and it is inferred that in W-containing steels, strength rises by suppressing recovery and re-crystallization of martensitic structures during creep.

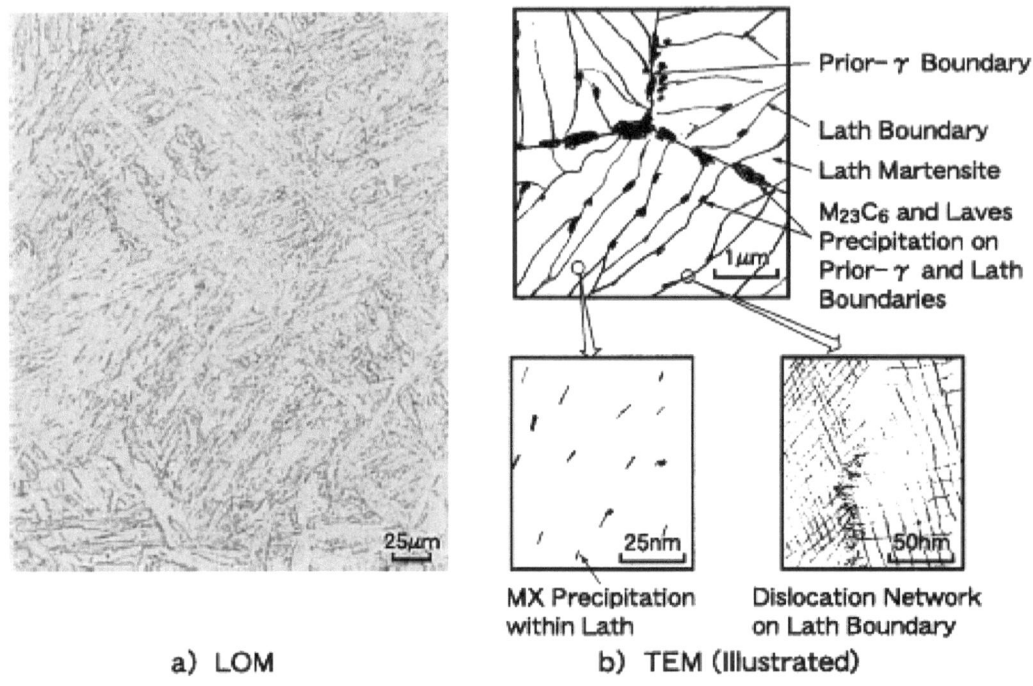

a) LOM   b) TEM (Illustrated)

Fig. 2.17. Typical microstructure of tempered martenstitc 9–12% Cr steel [21].

In order to facilitate a better understanding of the different types of steels, Figure 2.18 &2.19 shows schematically illustrated microstructures of ferritic and austenitic heat resistant materials. In both cases, material upgrades are illustrated from left to right, and the precipitates appearing therein change according to type [31].

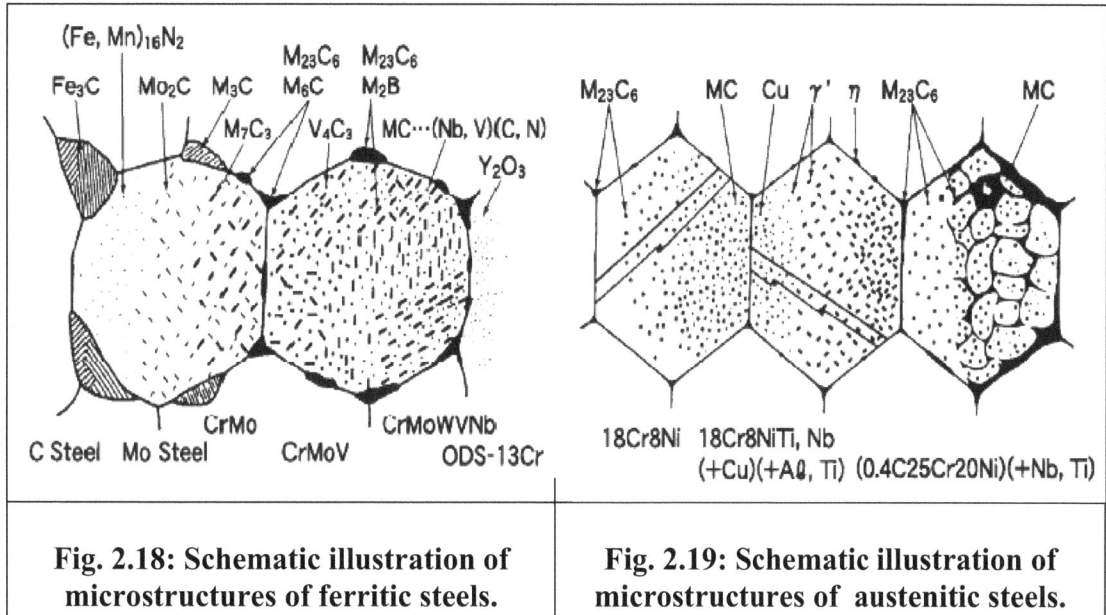

| Fig. 2.18: Schematic illustration of microstructures of ferritic steels. | Fig. 2.19: Schematic illustration of microstructures of austenitic steels. |

2. 3. 1. Comparison of the microstructure of P91, P92 and E911 :

High strength creep resistant steels are usually subjected to normalizing and tempering heat treatment prior to service and therefore, microstructure of those is tempered martensite. Creep strength of these steels are improved by its martensitic lath structure, precipitation strengthening effects of $M_{23}C_6$ carbide and MX (M= Nb, V, Cr and X= C, N) carbonitrides and soiled solution strengthening effect of Mo and W in the matrix. Especially, precipitation strengthening effect of MX carbonitrides is important because its coarsening rate is small and fine particles size is maintained for long-term [32].

P J Ennis1 and A Czyrska [33]; comparing the microstructure of the three types of steels (given in Table 4) after the standard heat treatment by normalized in the range 1040-1090 °C and Tempered in the range of 740-780 °C, all three steels exhibited similar microstructures, as shown in Figure 2.20.

Table 2.4: Details of high chromium steels, chemical compositions in wt% [33].

Element	P9	P91	P92	E911
C	max. 0.15	0.10	0.124	0.105
Si	0.20-0.65	0.38	0.02	0.20
Mn	0.80-1.30	0.46	0.47	0.35
P	max. 0.030	0.020	0.011	0.007
S	max. 0.030	0.002	0.006	0.003
Cr	8.5-10.5	8.10	9.07	9.16
Mo	1.70-2.30	0.92	0.46	1.01
W	-	-	1.78	1.00
V	0.20-0.40	0.18	0.19	0.23
Nb	0.30-0.45	0.073	0.063	0.068
B	-	-	0.003	-
N	-	0.049	0.043	0.072
Ni	max. 0.30	0.33	0.06	0.07
Al	-	0.034	0.002	-

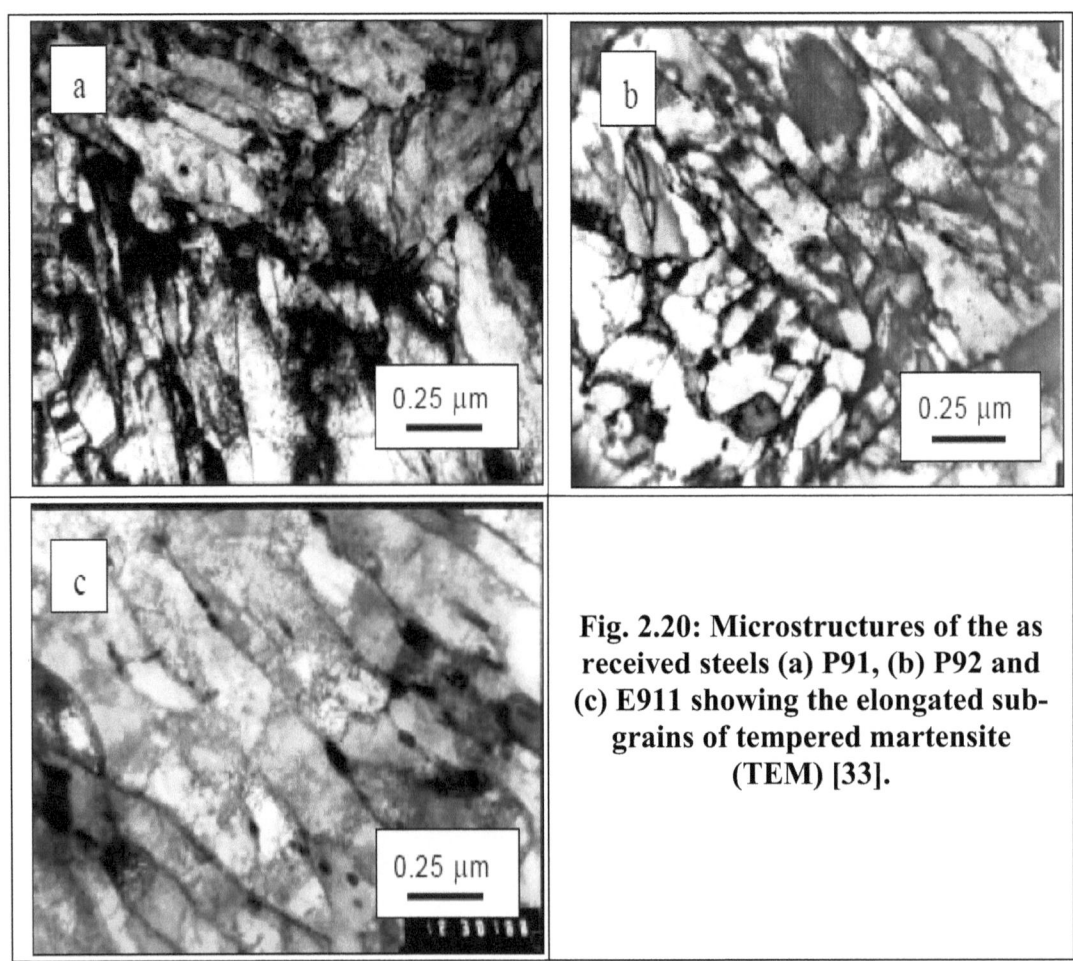

Fig. 2.20: Microstructures of the as received steels (a) P91, (b) P92 and (c) E911 showing the elongated sub-grains of tempered martensite (TEM) [33].

Austenitising produced a martensitic structure with a high dislocation density within the martensite laths. During tempering recovery caused the formation of sub-grains and dislocation networks. The creep strength of 9-12% Cr steels, is correlated inversely with the martensite lath width and therefore with the sub-grain size. Measurements of the average sub-grain width and of the dislocation density within the sub-grains, performed by means of quantitative TEM, are presented in Table 2.5.

Table 2.5 . Dislocation density & mean sub-grain size of as received P91, P92, E911 steels

Steel	Dislocation Density x 10^{-14}, m^{-2}	Mean Sub-Grain Size, µm
P91	7.5 ± 0.8	0.4 ± 0.06
P92	7.9 ± 0.8	0.4 ± 0.09
E911	6.5 ± 0.6	0.5 ± 0.05

It can be seen that sub-grain size is fairly similar in all steels investigated. The small differences are connected with different prior austenite grain size. Dislocation densities in P91 and P92 steels were similar, in both steels a little higher than in E911 steel. However, it must be borne in mind that only one heat of each steel has been examined in detail and the small differences observed may not be significant in the light of heat to heat variations. Besides recovery processes, the precipitation of carbides, carbonitrides and nitrides occurred during tempering. In all three steels examined, the $M_{23}C_6$ carbides containing Cr, Fe, Mo (W) precipitated preferentially on the prior austenite grain boundaries and on the martensite lath boundaries. These precipitates retard the sub-grain growth and therefore increase the strength of the material. In P91 steel mainly spheroidal Nb-rich carbonitrides are observed within the martensite laths. In the P92 and E911 steels three types of MX; Nb(C,N), plate-like VN and small complex Nb(C,N)-VN, were found [33].

A. Di Gianfrancesco and L. Cipolla; indicated the presence of small amout (< 3%) of δ-Ferrite in the microstructure of E911 steel, and they found that; the percentage of δ-Ferrite is a function of austenizing temperature and holding time and increase as they increase; started from 1060 °C and above [34]. J. Pasternak reported the presence of small amount of δ-Ferrite in the microstructure of P 92 steel [35].

2. 4. Microstructure stability and ageing effect :

The availability of long term creep tests is mandatory to perform reliable data assessment for rupture up to 10^5 hours and more, which are very important for the designer of the components for high temperature service. Advanced characterization of the microstructures for the high chromium content steels is necessary to confirm and guarantee that such steels are not affected by unexpected breakdowns of the creep properties due to microstructure instability.

Based on the above information of the previous paragraph comparing the initial microstructure of the three 9-12% Cr steels (P91, P92 and E911); we can summarize the initial microstructure of each steel through table 2.6 [33, 34 & 35]:

Table 2.6 : Initial microstructure of Heat treated (Normalized in the range 1040-1090°C and Tempered in the range of 740-780 °C) P91, P92 and E911 Steels.

Description	P91	P92	E 911
Lath of Martensite	X	X	X
Dislocation	X	X	X (Little Lower)
Sub-grains	X	X	X (Little wider)
Carbides	$M_{23}C_6$ carbides containing Cr, Fe, Mo	$M_{23}C_6$ carbides containing Cr, Fe, Mo (W)	$M_{23}C_6$ carbides containing Cr, Fe, Mo (W)
Carbonitrides	MX (Spheroidal Nb-rich)	MX; Nb(C,N), plate-like VN and small complex Nb(C,N)-VN	MX; Nb(C,N), plate-like VN and small complex Nb(C,N)-VN
δ-Ferrite	----	Small amount	< 3%
Lave Phase	-----	-----	-----

In the authors opinion [33, 34 & 35] there are strong indications that the controlling creep mechanism in the whole of the technically relevant stress/temperature range is dislocation creep. Microstructural sources of creep deformation are then the migration of dislocations and sub-grain boundaries. This indicates that microstructural explanations for the improved creep strength of the new 9-12%Cr steels should be found in mechanisms which retard the migration of dislocations and sub-grain boundaries, and thus delay the accumulation of creep strain with time. Such mechanisms include i) solid solution strengthening by formation of clouds of solute atoms around dislocations, and ii) interactions with precipitate particles [39].

Solid solution strengthening has often been referred to in discussions of the effect of Mo and W on creep strength of 9-12%Cr steels. It has since long been clear that during creep exposure at temperatures around 600°C-650°C most of the Mo and W in the steels will precipitate as intermetallic Laves phase ((Fe,Cr)2(Mo,W)). The dominating opinion was that this would cause creep instability in the steels because the solid solution strengthening effect from Mo and W would be lost, and the precipitation strengthening effect from Laves phase was believed to be insignificant. This opinion seemed to be supported by the breakdown of long-term creep strength of some W alloyed 9-11Cr steels, and by the excellent long-term stability of the Mo alloyed 9CrB turbine material Steel B from the European COST 501 project (Mo causes less Laves phase precipitation than W). However, the excellent long-term stability of the W alloyed 9Cr steel P92 seemed to contradict the opinion [39].

Recently, it has become clear that solid solution strengthening from W and Mo has no significant effect on long-term microstructure stability of the 9-12%Cr steels. Precipitation hardening by pinning of dislocations and sub-grain boundaries should be regarded as the most significant strengthening mechanism in 9-12%Cr steels, and microstructure stability of the 9-12%Cr steels under creep load is equivalent to precipitate stability [39]. This is consistent with two findings: a) the compositional changes, which have improved the creep strength of the 9-12%Cr steels, have also resulted in clear changes in the precipitate populations, and b) breakdowns in creep stability presented below can be explained by unexpected precipitate reactions.

F. Vodopivec *et al.* [36 & 37], found that; after normalizing and tempering no Laves phase is present in the material. However, during high temperature exposure, $M_{23}C_6$ carbides coarse, increase their mean dimensions; in the mean time precipitation and coarsening of intermetallic Laves phase $((Fe,Cr)_2(Mo,W))$ occurs; and as a result, the solid solution strengthening is reduced, due to the dissolution of elements like Mo, W and Cr.

The coarsening of precipitates with increasing ageing time and temperature is clearly shown in Figure 2.21, where the precipitate size distributions (Referring to MX, $M_{23}C_6$ and Laves precipitates), in five samples, submitted to different ageing conditions, are compared (As-treated, 550 °C/110,000h; 600 °C/58,439h; 650 °C/53,318h, characterized by different values of Larson-Miller Parameter, LMP).

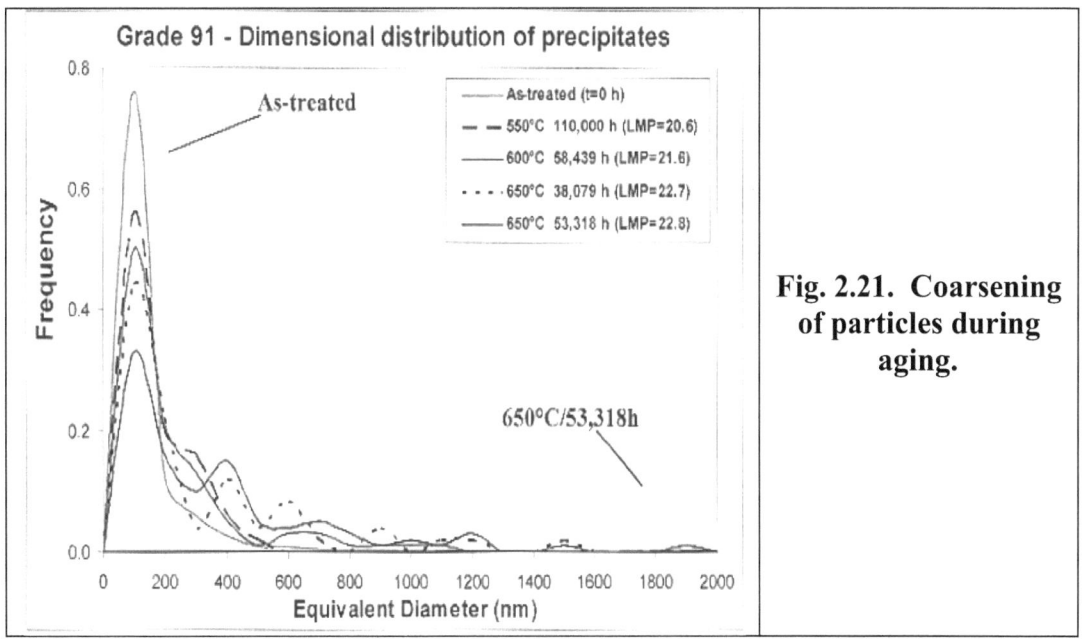

Fig. 2.21. Coarsening of particles during aging.

A summary of the precipitation evolution in grade 911 at 600 °C and 650 °C is shown in figures 2.12 and 2.13, respectively, in comparison with the growth of precipitate phases in Grade 91. The kinetics of $M_{23}C_6$ carbides and MX carbonitrides are very similar in both steels at 600 °C; no coarsening of MX carbonitrides was observed and the mean dimension of $M_{23}C_6$ carbides remained below 250 nm, even after a very long exposure and it is expected that they do not grow over. At 650 °C the coarsening of the $M_{23}C_6$ is much higher for grade 911 than for grade 91. Regarding to the laves phase the coarsening phenomena are similar at 600 °C, with a fast increase of the dimension at 650 °C for grade 911; which can be explained by a smaller inter-distance of Laves phase particles in Grade 911, due to a higher volume fraction and similar size respect to those observed in Grade 91, which accelerates the diffusion controlled growth.

J. Hald *et. al* [38 & 39]., reported the same when comparing steels P91 and P92 for long term exposure at 600 °C; in both steels the MX carbonitrides show a similar very high stability against coarsening. However, he reported also that; MX carbonitrides rich in V and Nb are extremely stable against coarsening, but they may be dissolved by precipitation of the complex nitride Z-phase (Cr(V,Nb)N). This is mainly a problem in Nb containing steels with Cr contents of 10% and above. The Z-phase nitride precipitate as very large particles, which will not contribute to precipitation hardening and which may each consume more than 1000 MX particles, Figure 2.22.

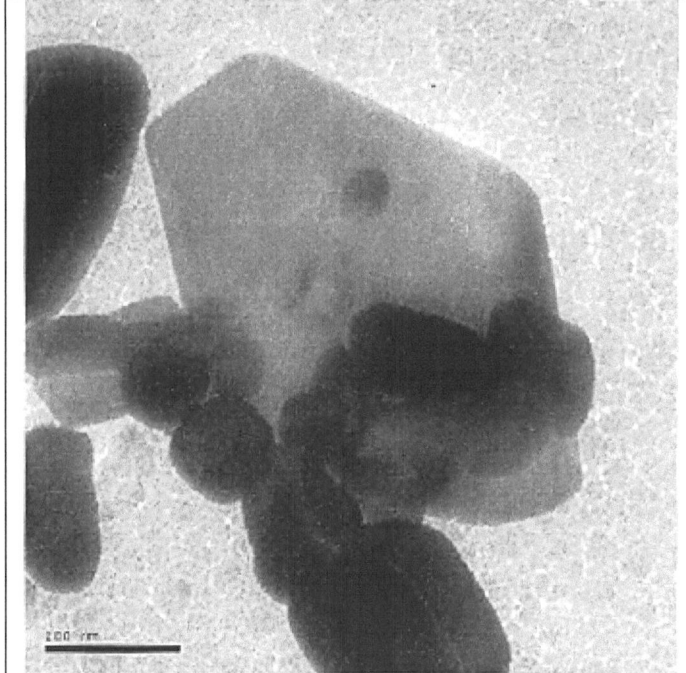

Fig. 2.22. Z-Phase Particle in a 11CrWCoVNbNB Steel.

Fujio *et. al*, reported that; the coarsening of $M_{23}C_6$ carbides in 12CrMo(W)VNbN steels is accompanied by dissolution of fine MX carbonitried due to precipitation of of coarse M_6X and or Z-Phases. Increase in Ni content in 12CrMoV steel results in accelerated microstructure degradation with more rapid coarsening of $M_{23}C_6$, dissolution of MX, precipitation M_6X and Fe_2Mo [40]. However, the $M_{23}C_6$ carbides show considerable coarsening in steel P91, whereas in steel P92 this carbide is highly stable. This is attributed to the presence of B in steel P92, which seem to improve the carbide coarsening stability [39]. On the other hand, large differences are found between the Laves phase particle sizes in the two steels. W produces fine stable (against coarsening) Laves phase particles in steel P92-even finer than the $M_{23}C_6$ carbides in steel P91- when the creep temperature is sufficiently below the solubility temperature for this phase. The observed difference in Laves phase behavior in steels P91 and P92 is explained by the lower solubility of the Mo Laves phases (650 °C) with W Laves phases (720 °C). When the creep temperature is closed to the solution temperature for Laves phase, nucleation of the phase is very difficult, and only few particles nucleate. This results in extended growth phase and large mean particle size. In fact, the observed differences in particle stability can explain the observed differences in creep rupture strength between the steels. The Laves phases in steel P92 produce a significant precipitate strengthening effect, which together with the observed high stability of $M_{23}C_6$ carbides provide the high creep strength of this steel [39].

Finally, as mentioned above; detailed microstructure studies have improved the understanding of the long term stability of the 9-12%Cr steels: Precipitation hardening controls the long-term microstructure stability, and solid solution strengthening from Mo and W plays no significant role in the long-term microstructure stability of 9-12%Cr steels. Significant particle strengthening can be obtained by intermetallic Laves phases provided that a large number of particles nucleate during creep exposure. MX carbonitrides rich in V and Nb are extremely stable against coarsening, but they may be dissolved by precipitation of the complex nitride Z-phase (Cr(V,Nb)N). This is mainly a problem in Nb containing steels with Cr contents of 10% and above. Significant advances in microstructure stability modeling will allow further improvements of the creep properties of this class of alloys.

2. 5. Understanding of main characteristics of creep resistant steels:

When selecting materials for high temperature service, several factors must be considered such as material cost, density, resistance to environmental attack under normal operating conditions and ability to resist serious distortion or failure during service. When assessing the resistance of materials to deformation and failure over long times under load at high temperature, attention must be given to the phenomenon of creep. Creep is often important in engineering design and is particularly relevant in applications involving high temperatures, such as steam turbines, piping within power plants, jet and rocket engines and nuclear reactor.

2. 5. 1. Creep and stress rupture strength:

At high temperatures (Prevalent at about half the melting point (0.4 ^T) and above), metals undergo time-dependent plastic straining when loaded. This phenomenon is known as creep and it can occur at stress levels less than the yield strength. Extension of an involved component may eventually produce a troublesome loss of dimensional tolerance or even ultimately lead to catastrophic rupture. Turbine blade creep due to the inhospitable temperatures of a jet engine is an often-cited example. High-pressure boilers and steam lines and nuclear reactor fuel cladding are components and systems that are also susceptible to creep effects [42].

Creep tests are performed on round tensile like specimens that are stressed by fixed suspended loads, while being heated by furnaces that coaxially surround them. The typical response obtained in a creep test is shown in Figure 2.23, where the specimen elongation or strain is recorded as a function of time. Initial elastic extension or strain started with instant applied load. Then a viscous like plastic straining ensues in which the creep strain rate ($\dot{\varepsilon} = d\varepsilon/dt$) decreases with time. This primary creep period then merges with the secondary creep stage where the strain rate is fairly constant. Alternately known as steady-state creep, this region of minimum creep normally occupies most of the test lifetime. The specimen undergoes a viscous like extension in this range. Finally the strain rate increases rapidly in the tertiary creep stage, leading to rupture of the specimen[42&43].

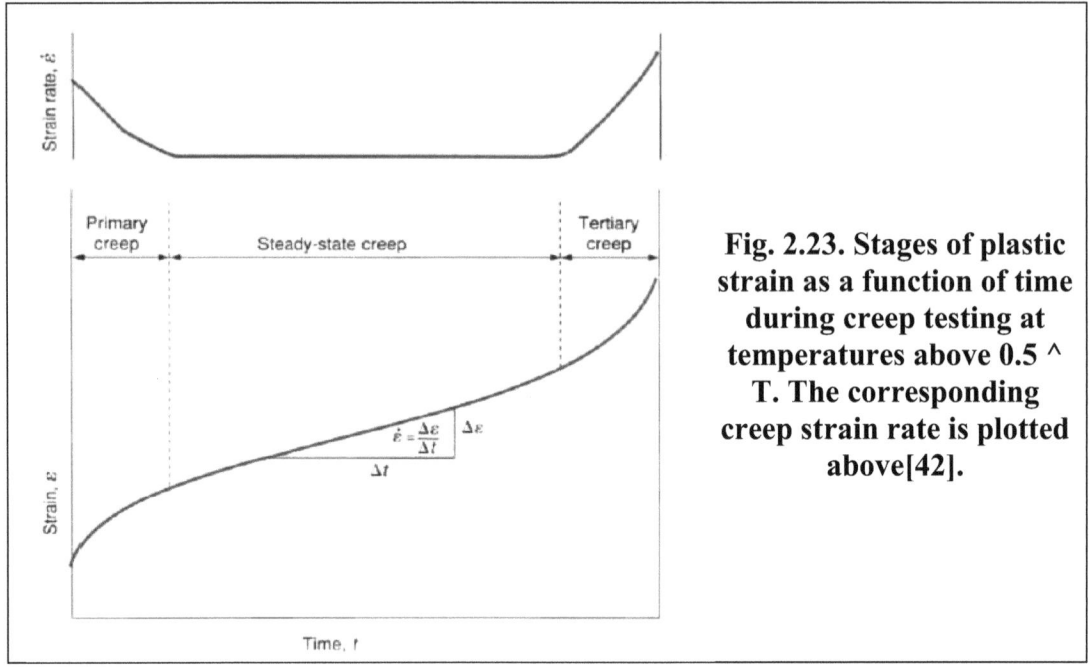

Fig. 2.23. Stages of plastic strain as a function of time during creep testing at temperatures above 0.5 ^ T. The corresponding creep strain rate is plotted above[42].

A family of curves can be produced which show the creep for different initial stresses and temperatures (Figure 2.24). At low stress and/or low temperature (curve I in the figure) *some primary* creep may occur but this falls to a negligible amount in the *secondary* stage when the creep curve becomes almost horizontal. With increased stress and/or temperature (curves II and III) the rate of secondary creep increases, leading to *tertiary* creep and inevitable catastrophic failure[43&46].

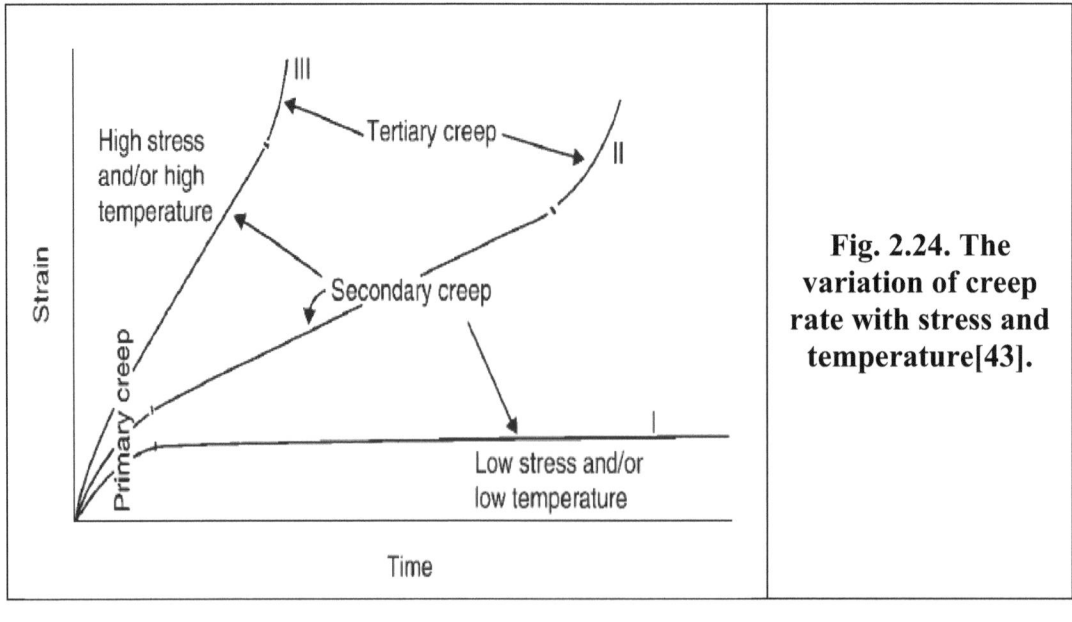

Fig. 2.24. The variation of creep rate with stress and temperature[43].

Creep in its simplest form is the progressive accumulation of plastic strain in a specimen or machine part under stress at elevated temperature over a period of time. Creep failure occurs when the accumulated creep strain results in a deformation of the machine part that exceeds the design limits. *Creep rupture* is an extension of the creep process to the limiting condition where the stressed member actually separates into two parts. *Stress rupture* is a term used interchangeably by many with creep rupture; however, others reserve the term stress rupture for the rupture termination of a creep process in which steady-state creep is never reached, and use the term creep rupture for the rupture termination of a creep process in which a period of steady-state creep has persisted. Figure 25 illustrates these differences. The interaction of creep and stress rupture with cyclic stressing and the fatigue process has not yet been clearly understood but is of great importance in many modern high-performance engineering systems. Creep strains of engineering significance are not usually encountered until the operating temperatures reach a range of approximately 35–70% of the melting point on a scale of absolute temperature [44].

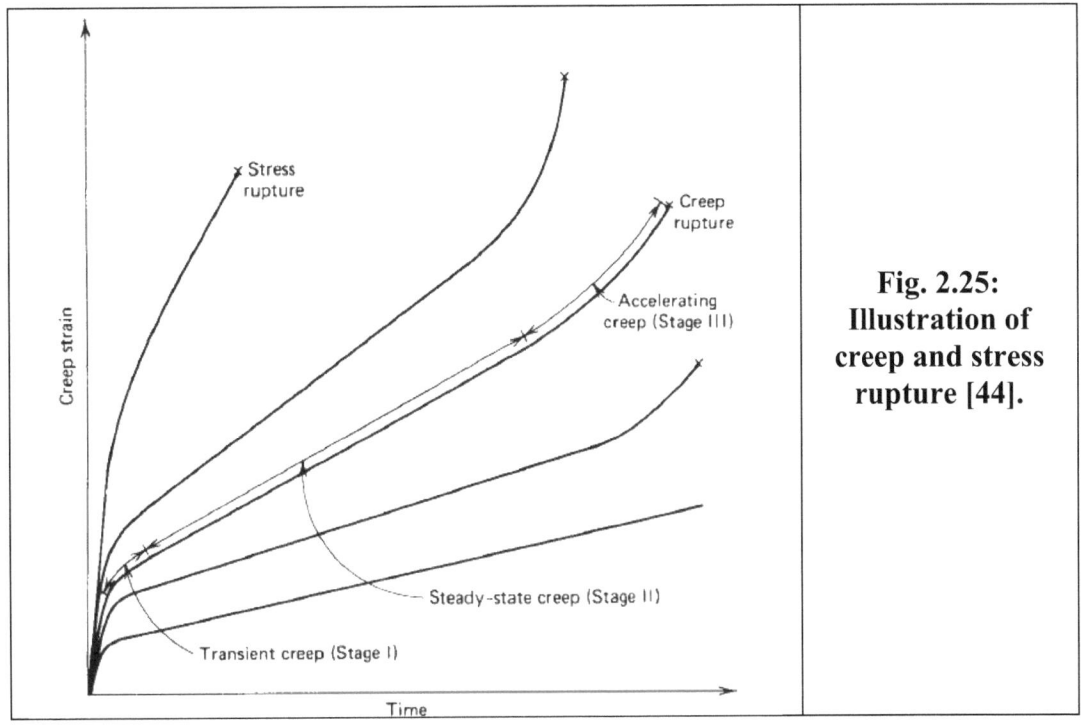

Fig. 2.25: Illustration of creep and stress rupture [44].

Not only is excessive deformation due to creep an important consideration, but other consequences of the creep process may also be important. These might include creep rupture, thermal relaxation, dynamic creep under cyclic loads or cyclic temperatures, creep and rupture under multiaxial states of stress, cumulative creep effects, and effects of combined creep and fatigue.

Creep deformation and rupture are initiated in the grain boundaries and proceed by sliding and separation. Thus, creep rupture failures are intercrystalline, in contrast, for example, to the transcrystalline failure surface exhibited by room temperature fatigue failures. Although creep is a plastic flow phenomenon, the intercrystalline failure path gives a rupture surface that has the appearance of brittle fracture.

Creep rupture typically occurs without necking and without warning. Current state-of-the-art knowledge does not permit a reliable prediction of creep or stress rupture properties on a theoretical basis. Furthermore, there seems to be little or no correlation between the creep properties of a material and its room temperature mechanical properties. Therefore, test data and empirical methods of extending these data are relied on heavily for prediction of creep behavior under anticipated service conditions. Metallurgical stability under long-time exposure to elevated temperatures is mandatory for good creep-resistant alloys. Prolonged time at elevated temperatures acts as a tempering process, and any improvement in properties originally gained by quenching may be lost. Resistance to oxidation and other corrosive media are also usually important attributes for a good creep-resistant alloy. Larger grain size may also be advantageous since this reduces the length of grain boundary, where much of the creep process resides[44].

To better understand the creep response of materials, engineers perform two types of tests. The first aims to determine the steady-state creep rate over a suitable matrix of stress and temperature. Thus tests are performed at the *same stress* but at *different temperatures,* as well as at the *same temperature* but at *different stresses,* as shown in Figure 2.26. Specimens are usually not brought to failure in such tests; accurately predicting their extension is of interest here. The second, known as the creep rupture test, is conducted at higher stress and temperature levels to accelerate failure. Such test information can either be used to estimate short-term life (e.g., turbine blades in military aircraft) or be extrapolated to lower service temperatures and stresses (e.g., turbine blades in utility power plants) to predict long-term life[42].

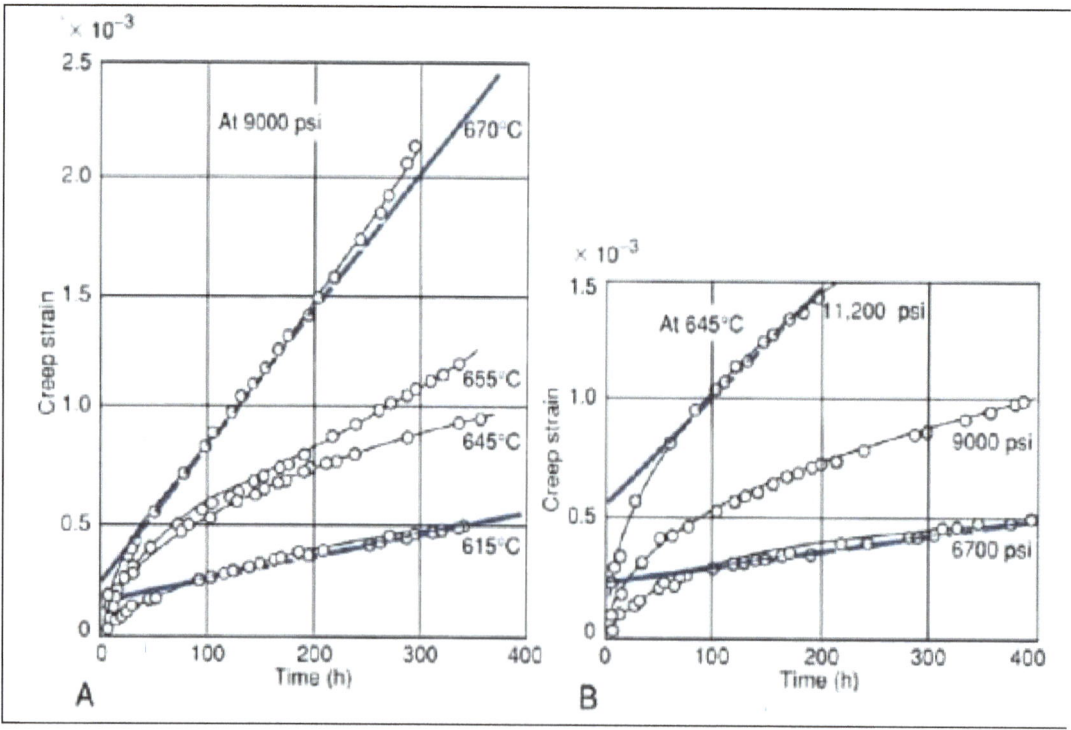

Fig. 2.26. Creep strain versus time test results in a 0.5 wt% Mo, 0.23 wt% V steel. (A) Constant stress, variable temperature. (B) Constant temperature, variable stress [42].

Creep tests are carried out on test-pieces which are similar in form to ordinary test-pieces. A test-piece is enclosed in a thermostatically controlled electric tube furnace which can be maintained accurately at a fixed temperature over the long period of time occupied by the test. The test-piece is statically stressed, and some form of sensitive extensometer is used to measure the extremely small extensions at suitable time intervals. A set of creep curves, obtained for different static forces at the same temperature, is finally produced, and from these the limiting creep stress is derived.

The stress which will produce a limiting amount of creep in a fixed time, say two or three days, at a particular temperature can be determined. Thus the stress/temperature combination which will produce a limiting value of creep strain of, say, 10~3 in a time of 1000 hours might be determined. The term *stress to rupture* of a material is used for the initial stress which will result in failure at a particular temperature after some period of time. For example, for a 0.2 per cent plain carbon steel at 400°C, the stress to rupture in 1000 hours is 295 MPa; at 500°C it has dropped to 118 MPa [43].

Stress-rupture is the "brute force method" in which a large number of tests are run at various stresses and temperatures to develop plots of applied stress vs. time to failure as shown in Figure 2.27. While it is relatively easy to use these plots to provide estimates of stress rupture life within the range of stresses and lives covered by the test data, extrapolation of the data can be problematic when the failure mechanism changes as a function of time or stress as shown by the "knee" in Figure 2.27[45].

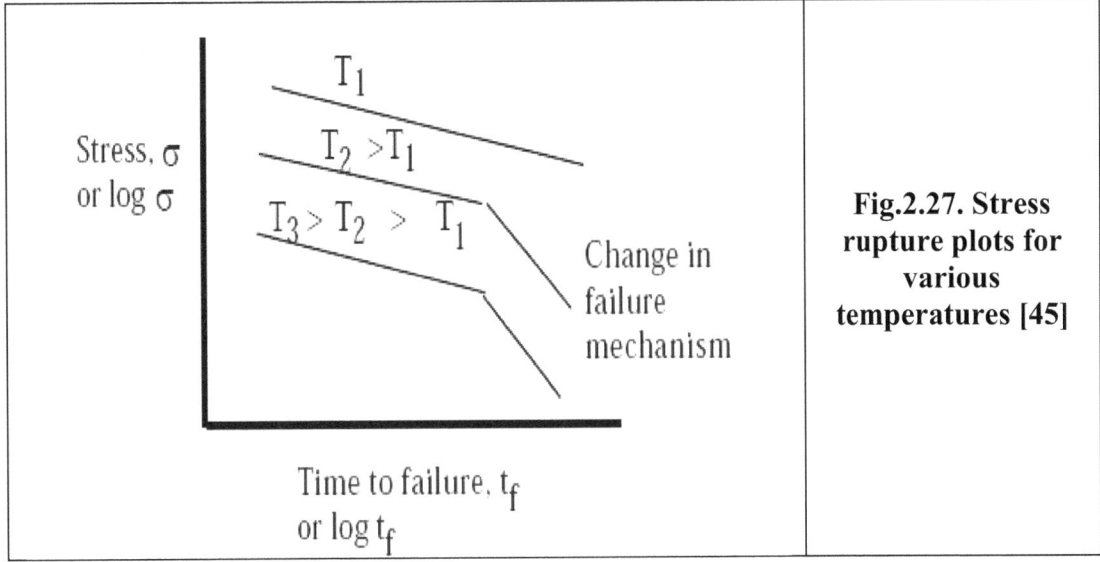

Fig.2.27. Stress rupture plots for various temperatures [45]

2. 5. 1. 1. Prediction of long-term creep behavior:

Much time and effort has been expended in attempting to develop good short time creep tests for accurate and reliable prediction of long-term creep and stress rupture behavior. It appears, however, that really reliable creep data can be obtained only by conducting long-term creep tests that duplicate actual service loading and temperature conditions as nearly as possible. Unfortunately, designers are unable to wait for years to obtain design data needed in creep failure analysis. Therefore, certain useful techniques have been developed for approximating long-term creep behavior based on a series of short-term tests. Data from creep testing may be plotted in a variety of different ways. The basic variables involved are stress, strain, time, temperature, and, perhaps, strain rate.

Any two of these basic variables may be selected as plotting coordinates, with the remaining variables treated as parametric constants for a given curve. Three commonly used methods for extrapolating short-time creep data to long-term applications are the abridged method, the mechanical acceleration method, and the thermal acceleration method. In the abridged method of creep testing the tests are conducted at several different stress levels and at the contemplated operating temperature. The data are plotted as creep strain versus time for a family of stress levels, all run at constant temperature. The curves are plotted out to the laboratory test duration and then extrapolated to the required design life. In the mechanical acceleration method of creep testing, the stress levels used in the laboratory tests are significantly higher than the contemplated design stress levels, so the limiting design strains are reached in a much shorter time than in actual service. The data taken in the mechanical acceleration method are plotted as stress level versus time for a family of constant strain curves all run at a constant temperature. The thermal acceleration method involves laboratory testing at temperatures much higher than the actual service temperature expected. The data are plotted as stress versus time for a family of constant temperatures where the creep strain produced is constant for the whole plot. It is important to recognize that such extrapolations are not able to predict the potential of failure by creep rupture prior to reaching the creep design life [44].

2. 5. 1. 1. 1. Extrapolation schemes :

In any testing method it should be noted that creep testing guidelines usually dictate that test periods of less than 1% of the expected life are not deemed to give significant results. Tests extending to at least 10% of the expected life are preferred where feasible[44]. Several different theories have been proposed to correlate the results of short time elevated-temperature tests with long-term service performance at more moderate temperatures. Often tests are limited to 1000 h (42 days). To ensure that neither rupture nor excessive creep strain occurs, results from shorter time tests at higher temperatures and/or higher stresses must be extrapolated to the service conditions. Several schemes have been proposed for such extrapolation [46].

One of the more accurate and useful of these proposals is the Larson–Miller theory. The Larson–Miller theory postulates that for each combination of material and stress level there exists a unique value of a parameter P that is related to temperature and time by the equation:

$$P = (\theta + 460)(C + \log_{10} t) \quad \text{-------------- (1)}$$

where P = Larson–Miller parameter, constant for a given material and stress level
θ = temperature, °F
C = constant, usually assumed to be 20
t = time in hours to rupture or to reach a specified value of creep strain

This equation was investigated for both creep and rupture for some 28 different materials by Larson and Miller with good success. By using (Larson-Miller Equation) it is a simple matter to find a short-term combination of temperature and time that is equivalent to any desired long-term service requirement. For example, for any given material at a specified stress level the test conditions listed in Table 2.7 should be equivalent to the operating conditions.

Table 2.7 : Equivalent Conditions Based on Larson–Miller Parameter [44].

Operating Condition	Equivalent Test Condition
10,000 h at 1000°F	13 h at 1200°F
1,000 h at 1200°F	12 h at 1350°F
1,000 h at 1350°F	12 h at 1500°F
1,000 h at 300°F	2.2 h at 400°F

In this method, a number of tests are run at various temperatures and stresses to determine the times to failure and activation energy. A "universal" plot (see Figure 2.28) is then made of the stress as a function of P_{LM}. The allowable stress for an combination of time to failure and temperature (i.e., P_{LM}) can then determined from the curve[45].

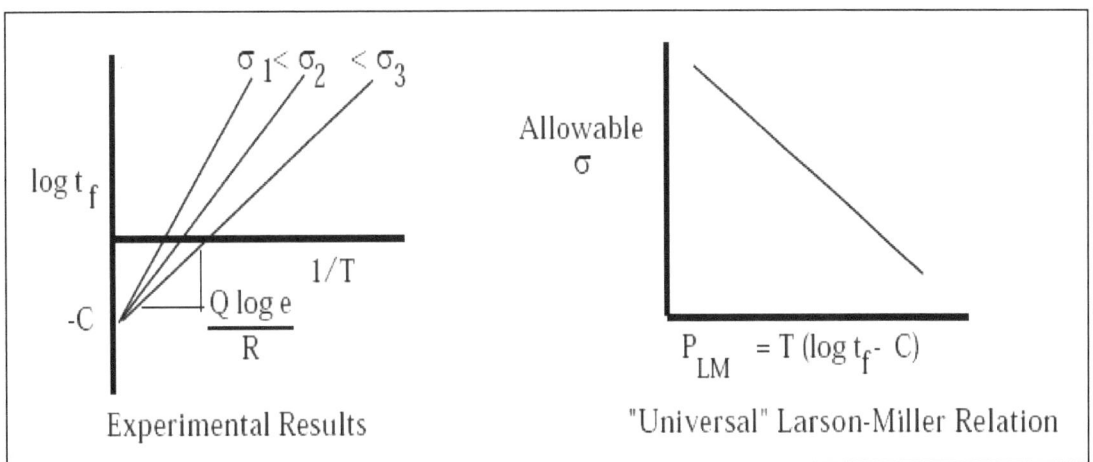

Fig. 2.28. Summary of Larson-Miller relation [45].

Where P_{LM} is the Larson-Miller parameter, Q is assumed to a function of stress only, and C is a constant of ~20 for most materials.

On the other hand, in the *Sherby-Dorn method*, θ is the temperature compensated time such that:

$$P_{SD} = \log \theta = \log t_f - \frac{\log e}{R} \frac{Q}{T} \qquad (2)$$

Where P_{SD} is the Sherby-Dorn parameter and Q is assumed independent of temperature and stress. In this method, a number of tests are run at various temperatures and stresses to determine the times to failure and activation energy. A "universal" plot (see Figure 2.29) is then made of the stress as a function of P_{SD}. The allowable stress for an combination of time to failure and temperature (i.e., P_{SD}) can then be determined from the curve[45].

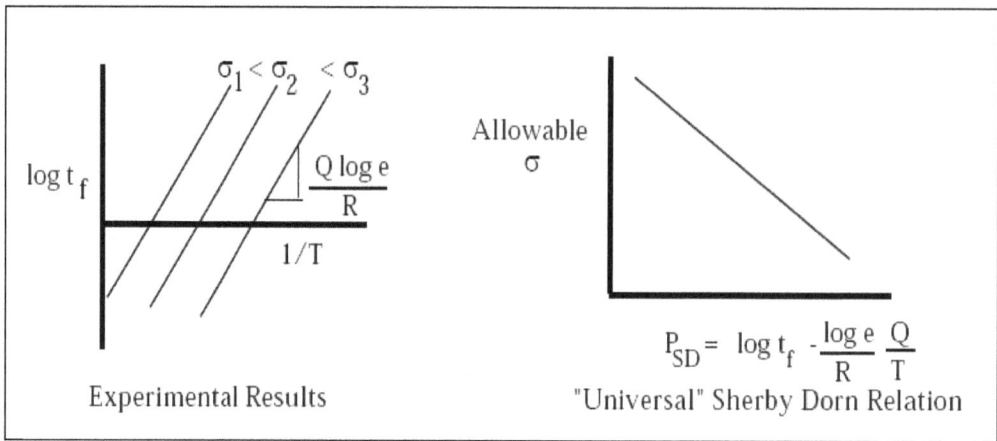

Fig. 2.29. Summary of Sherby-Dorn relation[45].

2. 5. 2. Techniques of creep testing and assessment of creep strength:

The creep test is conducted using a tensile specimen to which a constant stress is applied, often by the simple method of suspending weights from it. Surrounding the specimen is a thermostatically controlled furnace, the temperature being controlled by a thermocouple attached to the gauge length of the specimen. The extension of the specimen is measured by a very sensitive extensometer since the actual amount of deformation before failure may be only two or three per cent. The results of the test are then plotted on a graph of strain versus time which is known as the creep curve.

The test specimen design is based on a standard tensile specimen. It must be proportional in order that results can be compared and ideally should be machined to tighter tolerances than a standard tensile test piece. In particular the straightness of the specimen should be controlled to within some 0.5 % of the diameter. A slightly bent specimen will introduce bending stresses that will seriously affect the results.

Conceptually a creep test is rather simple: Apply a force to a test specimen and measure its dimensional change over time with exposure to a relatively high temperature. If a creep test is carried to its conclusion (that is, fracture of the test specimen), often without precise measurement of its dimensional change, then this is called a stress rupture test (see Figure 30). Although conceptually quite simple, creep tests in practice are more complicated. Temperature control is critical (fluctuation must be kept to <0.1 to 0.5°C). Resolution and stability of the extensometer is an important concern (for low creeping materials, displacement resolution must be on the order of 0.5 μm). Environmental effects can complicate creep tests by causing premature failures unrelated to elongation and thus must either mimic the actual use conditions or be controlled to isolate the failures to creep mechanisms. Uniformity of the applied stress is critical if the creep tests are to interpreted [45]. Figure 2.30 Comparison of creep and stress rupture tests

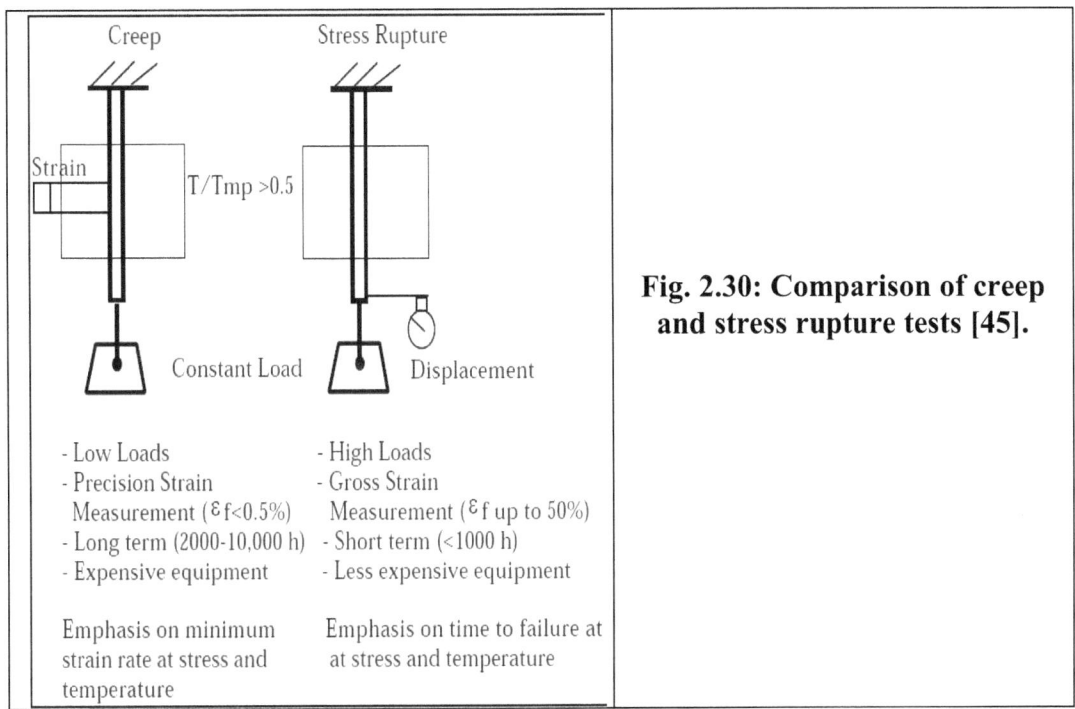

Fig. 2.30: Comparison of creep and stress rupture tests [45].

The chart in Figure 2.31 illustrates the most currently used creep testing methods [62]. These methods are briefly described as follows:

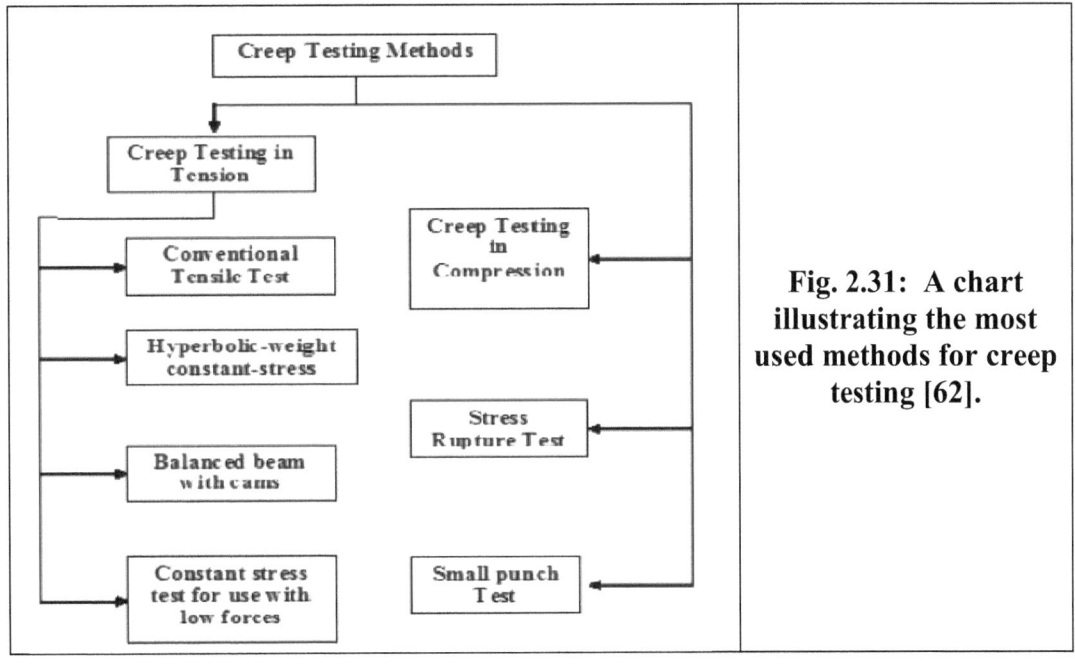

Fig. 2.31: A chart illustrating the most used methods for creep testing [62].

2. 5. 2. 1. Conventional tensile creep testing:

In this type of creep testing, a constant load is applied to a tensile specimen maintained at a constant temperature. Strain is then measured over a period of time. The creep test is usually employed to obtain the minimum creep rate of the tested material in Stage II from the slope of the creep curve at this stage. Engineers need to account for this expected deformation when designing systems. Creep tests in practice are more complicated. Temperature control is critical (fluctuation must be kept to (0.1 to 0.5 °C). Resolution and stability of the extensometer is an important concern (for low creeping materials, displacement resolution must be on the order of 0.5 µm). Environmental effects can complicate creep tests by causing premature failures unrelated to elongation and thus must either mimic the actual use conditions or be controlled to isolate the failures to creep mechanisms. Uniformity of the applied stress is critical if the creep tests are to interpreted [62]. Figure 2.32 shows a Typical creep test set-up that can be used for both creep and stress rupture tests [45].

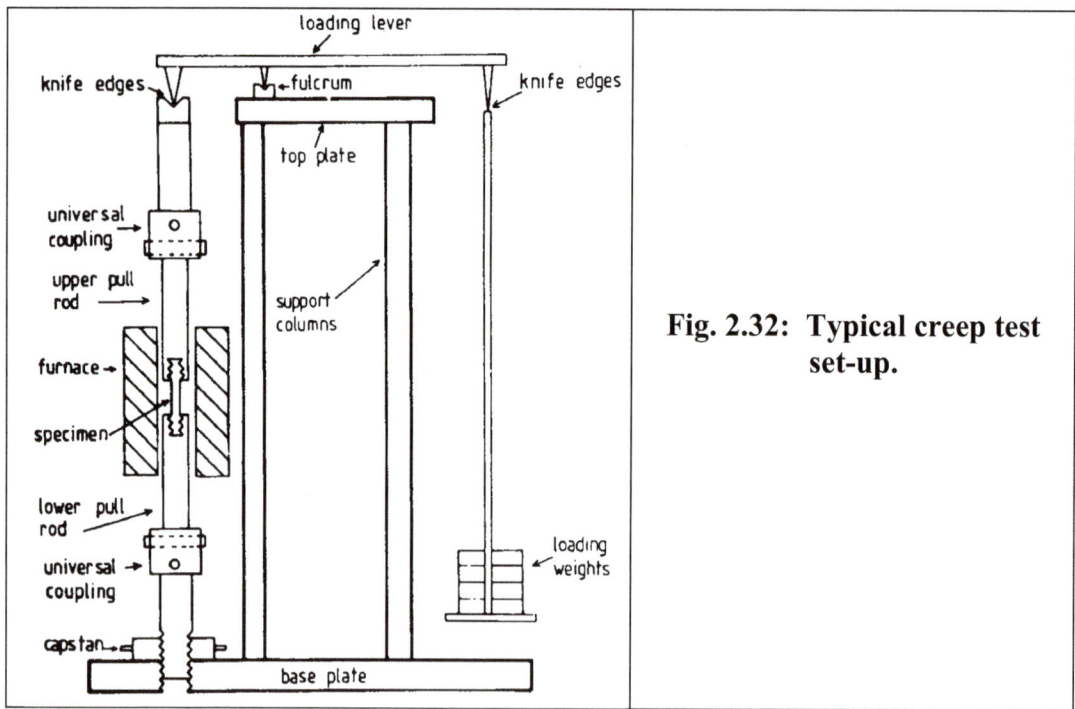

Fig. 2.32: Typical creep test set-up.

2. 5. 2. 2. Hyperbolic-weight constant-stress method:

An early application of this method in creep test was the use of a hyperbolic weight by Andrade in which load was reduced with extension of the specimen as the weight was lowered into a liquid (Figure 2.33). The required shape of the weight is given by an equation of a hyperbola:

$$y = \sqrt{\frac{M L_0}{\rho \pi}} \cdot \frac{1}{L_0 + x} \quad \text{------------- (3)}$$

Where M is the mass of the load, L_0 is the initial length of the wire, and ρ is the density of the liquid which is usually water [63].

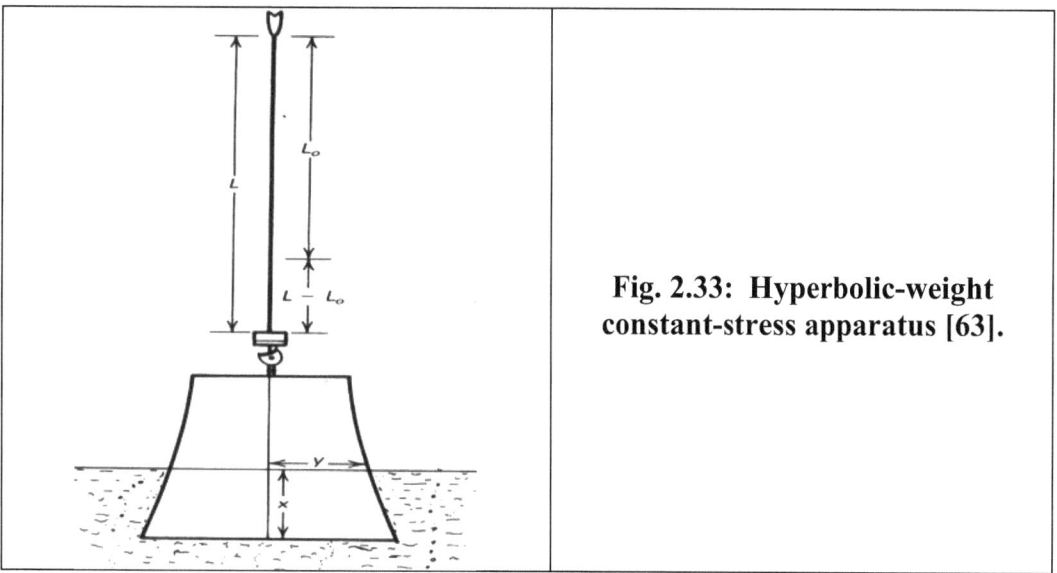

Fig. 2.33: Hyperbolic-weight constant-stress apparatus [63].

2. 5. 2. 3. Balanced beam with cams:

This is more convenient and useful method that uses a balanced beam with shaped cams in which a beam PH is supported by a knife B and carries two plates F and C, one at each end as shown in Figure 2.34 . The plate C has a grove along its outer edge HK; the profile HK is an arc of a circle of center B. D is a thin steel wire resting in a grove that is fixed to the adjusting screw E. The lower end of D is attached to the upper end of the wire to be attached. F is the second plate in the grove of which lies a thin steel wire supporting a weight W. The profile of the bottom of the grove PQR is made such that the moment of the weight W about the axis trough B is inversely proportional to length of the wire undergoing stretch, which will make the stretching force proportional to the cross section of the wire [63].

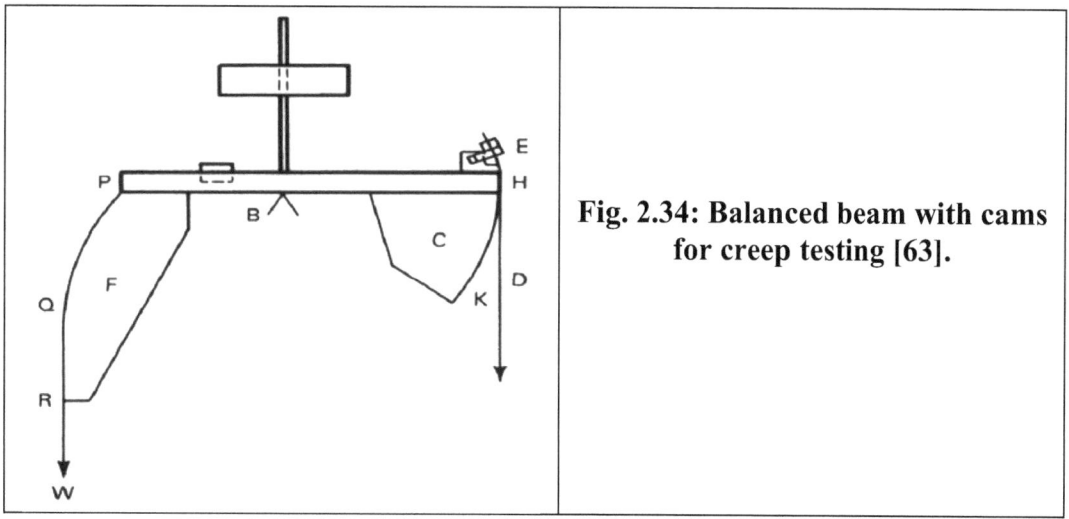

Fig. 2.34: Balanced beam with cams for creep testing [63].

2. 5. 2. 4. Constant stress test for use with low forces:

This type of testing has been developed to eliminate the frictional effect of hinges or bearings. Modifications of this type (such as the one shown in Figure 2.35) were designed to achieve the accurate maintenance of constant stress when forces as low as 0.1 N are involved. To balance the mass of the cam about the point of the suspension and thereby ensure that the only force on the specimen is the applied force due to the mass M, the counter-mass C is necessary. With the modified suspension, the cam profile equation becomes:

$$r = \frac{(r_0 - R)L_0}{L_0 + 2R(\theta - \theta_0)} + R \quad \text{---------------- (4)}$$

For the apparatus shown in Figure 2.35, the lower limit is 0.1 N, which corresponds to 5.0 KPa if the specimen diameter is 5 mm. Thus, the apparatus is suitable for use at small stresses, such as may be encountered in studies of plastic flow in temperatures close to the melting point [63].

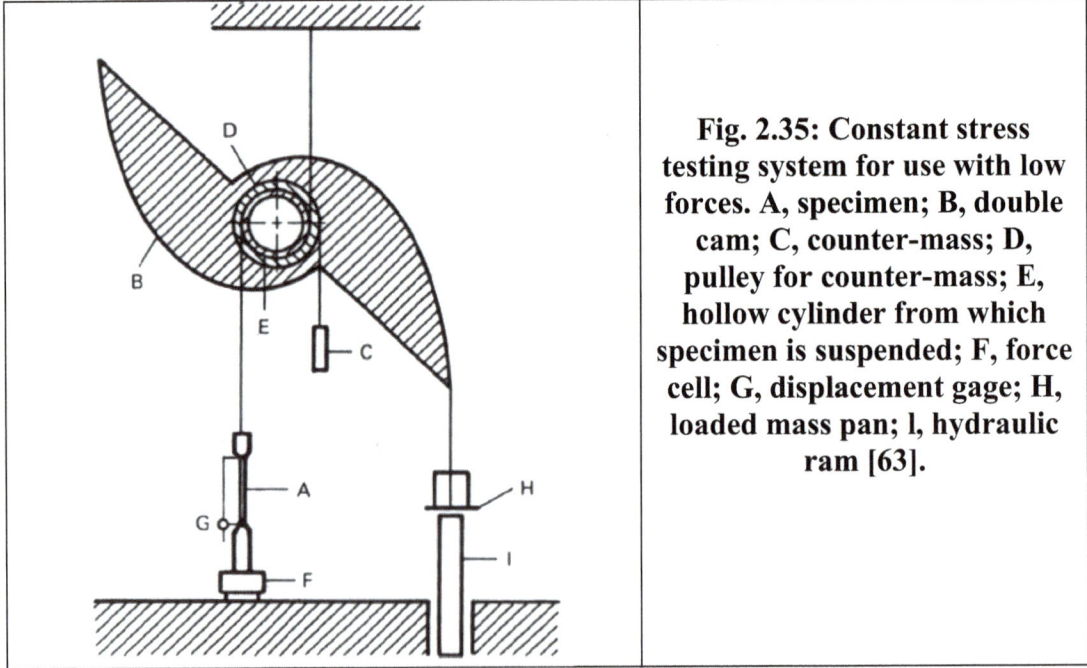

Fig. 2.35: Constant stress testing system for use with low forces. A, specimen; B, double cam; C, counter-mass; D, pulley for counter-mass; E, hollow cylinder from which specimen is suspended; F, force cell; G, displacement gage; H, loaded mass pan; I, hydraulic ram [63].

2. 5. 2. 5. High-temperature constant-stress compression creep testing:

The apparatus shown in Figure 2.36 is used for this type of creep testing. The following equation ensures maintenance of constant stress:

$$r_3 = \frac{L_1 L_2}{L} \quad \text{---------------- (5)}$$

Where L1, L2, and L3 are the distance between the various knife edges, and L is the sample length. Two knife edges provide the fulcrum for the lower arm, another supports the weight pan, and a fourth applies the force to the lower ample push rod. When the load pan support columns are vertical, the frame acts as a simple fulcrum to apply a force on the lower push rod that is greater than the force on the weight pan by a factor of L2/L1. When the sample deforms, the frame tilts backward, and the columns are no longer vertical. This movement which is proportional to the present L3 and the amount of deformation increases the force applied to the pushing rod and also to the sample, the increase in force compensates for the increase in area of the sample as it is compressed, thus maintaining a constant stress. Shorter samples require a greater L3 because the area of the sample becomes greater for the same amount of deformation than for a sample of greater height [63].

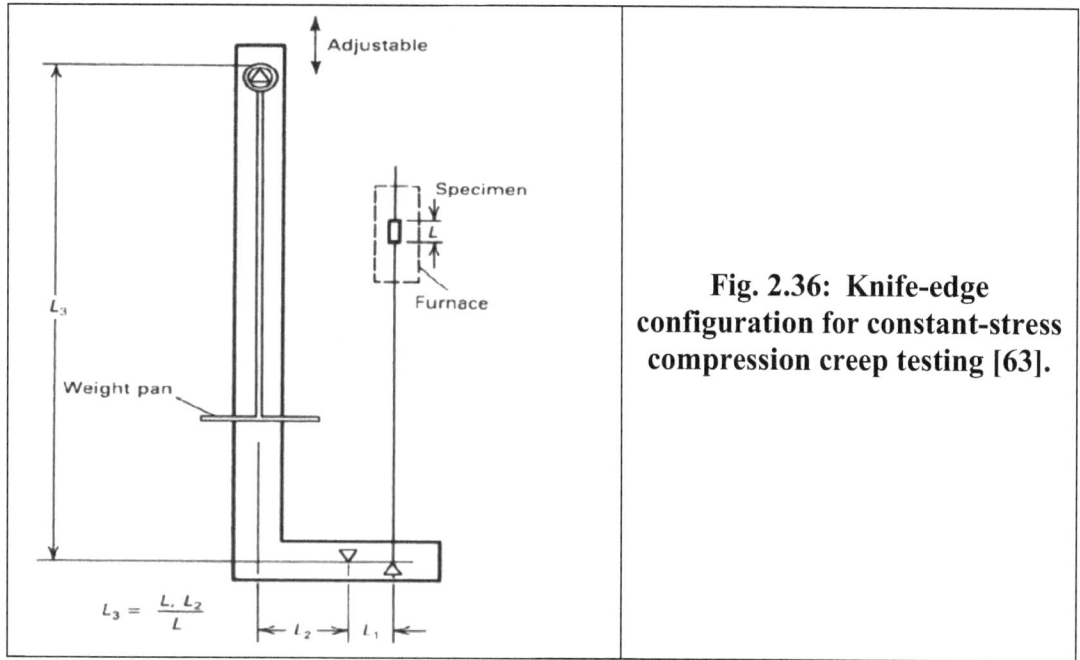

Fig. 2.36: Knife-edge configuration for constant-stress compression creep testing [63].

2. 5. 2. 6. Small-punch creep test:

The small-punch creep test (which is also known as the miniaturized disk-bend creep test) is considered nowadays in many testing laboratories all over the world as a short term creep testing technique for two main reasons, on one is that it reduces the time for many orders to assist the creep properties of the tested materials, and second is the using of too small testing samples which also reduces the cost of the material and sample preparation. This technique can be used through a short term investigation test to make a comparison of the creep strength of the different zones for a welded joint. Small-punch creep test was originally developed for estimating the fracture-appearance transition temperature of metallic materials. This test can also be used to acquire high-temperature creep data by deforming a supported disk-shaped test specimen with a spherical penetrator under constant load. The measured time dependence of the central deflection is closely related to the creep curves of conventional uni-axial creep test. Out of many testing techniques, small-punch test is currently used to investigate the mechanical properties such as creep strength of steels. In the small-punch test as it is shown in Figure 2.37, a thin circular disk is supported over a recessed hole and forced, under a constant load, to deform into the hole by means of a spherically shaped punch [64].

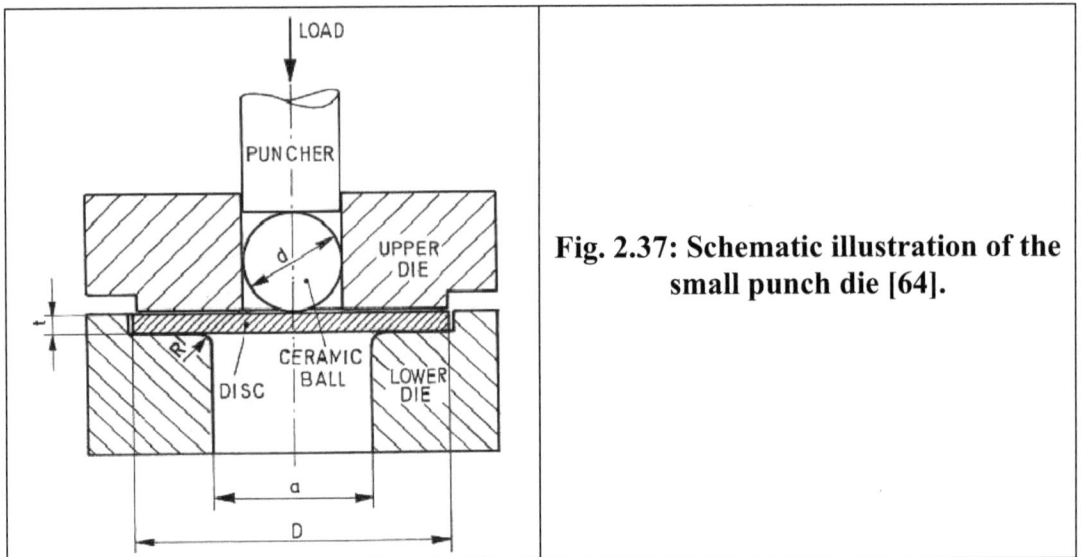

Fig. 2.37: Schematic illustration of the small punch die [64].

The central deflection (displacement) of the disk test specimen and its displacement rate are monitored as a function of time up to the point of failure of the disk. Initially the stresses in the disk are very high because at the beginning of the test, the load is transferred from the spherical surface of the punch as a point load onto the flat disk and these stresses are certainly higher than the corresponding yield stress of the steel at the test temperature. As a result, the disk buckles rapidly, and a large displacement of the punch occurs within a very short time.

Due to the buckling of the disk the size of the contact surface between the punch and the disk increases significantly, the stresses in the disk are reduced, and further deformation of the disk occurs only by creep. It has been demonstrated that the shape of the time dependence curve as measured from the central deflection of a disc-specimen is qualitatively similar to the shape of creep curves resulting from conventional uniaxial testing [64]. Small-punch test usually shows a load displacement curve with four distinct stages; elastic and plastic bending, plastic stretching and instability to fracture. However, the instability stage often disappears for disk bend tests on materials with high ductility; because the maximum strain available for this method is limited by the radius of the lower die [65].

2. 6. Creep damage and creep fracture:

During creep, damage, in the form of internal cavities, accumulates. The damage first appears at the start of the Tertiary Stage of the creep curve and grows at an increasing rate thereafter. The shape of the tertiary Stage of the creep curve (Figure 2.38) reflects this: as the cavities grow, the section of the sample decreases, and (at constant load) the stress goes up. Since the creep rate ε increases with σ_n, the creep rate goes up even faster than the stress does (Figure 2.38) [66].

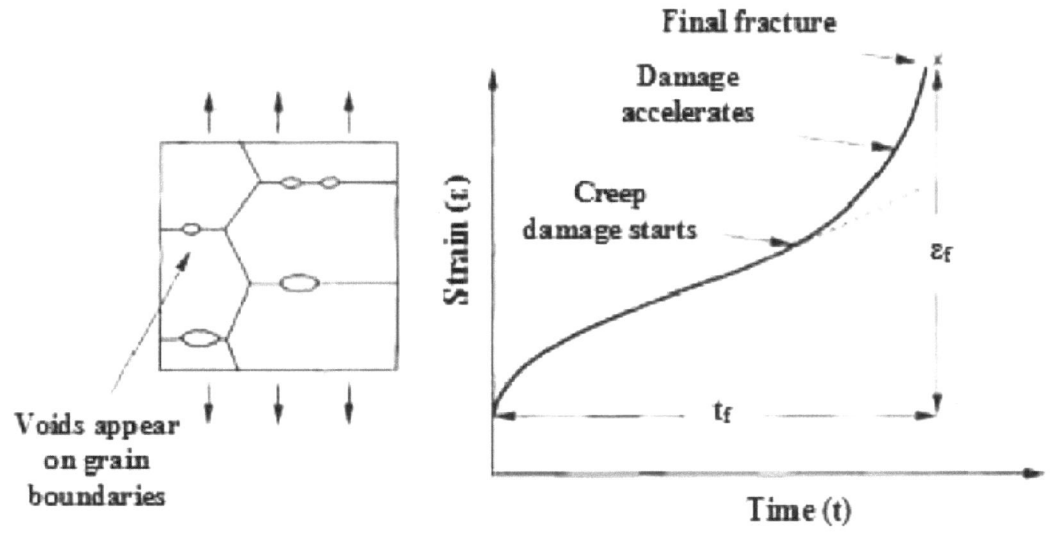

Fig. 2.38 : Creep damage in materials [66].

It is not surprising - since creep causes creep fracture - that the time-to-failure, t_f is described by a constitutive equation which looks very like that for creep itself:

$$t_f = A'\sigma^{-m} \exp\frac{Q}{RT} \quad \text{---------------- (6)}$$

Here A', m and Q are the creep-failure constants, determined in the same way as those for creep (the exponents have the opposite sign because t_f is a time whereas ε is a rate). In many high-strength alloys this creep damage appears early in life and leads to failure after small creep strains (as little as 1 %). In high-temperature design it is important to make sure:

(a) That the creep strain during the design life is acceptable
(b) That the creep ductility (strain to failure) is adequate to cope with the acceptable design life.
(c) That the time-to-failure at the design loads and temperatures is longer (by a creep strain suitable safety factor) than the design life [66].

2. 6. 1. Creep mechanisms of metals:
2. 6. 1. 1. Dislocation creep (giving power-law creep):

The stress required to make a crystalline material deform plastically is that needed to make the dislocations in it move. Their movement is resisted by, the intrinsic lattice resistance and the obstructing effect of obstacles (e.g. dissolved solute atoms, precipitates formed with un-dissolved solute atoms, or other dislocations). Diffusion of atoms can 'unlock' dislocations from obstacles in their path, and the movement of these unlocked dislocations under the applied stress is what leads to dislocation creep. How does this unlocking occur?, Figure 2.39 shows a dislocation which cannot glide because a precipitate blocks its path. The glide force τ_b per unit length is balanced by the reaction from the precipitate. But unless the dislocation hits the precipitate at its mid-plane (an unlikely event) there is a component of force left over [66].

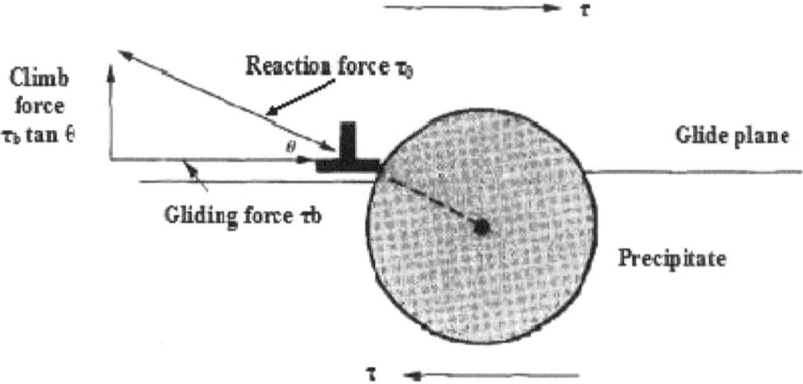

Fig. 39:. The climb force on a dislocation [66].

It is the component $\tau_b \tan \theta$ (θ is the inclination angle of the reaction force of the precipitate τ_0), which tries to push the dislocation out of its slip plane. The dislocation cannot glide upwards by the shearing of atom planes but the dislocation can move upwards if atoms at the bottom of the half-plane are able to diffuse away. A mechanical force can do exactly the same thing, and this is what leads to the diffusion of atoms away from the measuring of the creep property 'loaded' dislocation, eating away its extra half-plane of atoms until it can clear the precipitate. The process is called climb, and since it requires diffusion, it can occur only when the temperature is above 0.3 Tm or so (Figure 2.40) [66].

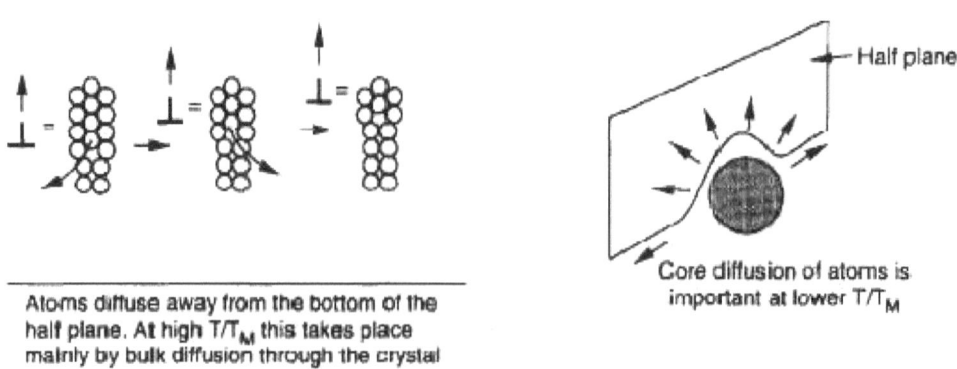

Fig. 2.40. How diffusion leads to climb [66].

Climb unlocks dislocations from the precipitates which pin them and further slip (or 'glide') can then take place (Figure. 2.41). Similar behavior takes place for pinning by solute, and by other dislocations. After a little glide, of course, the unlocked dislocations jump into the next obstacles, and the whole cycle repeats itself [66].

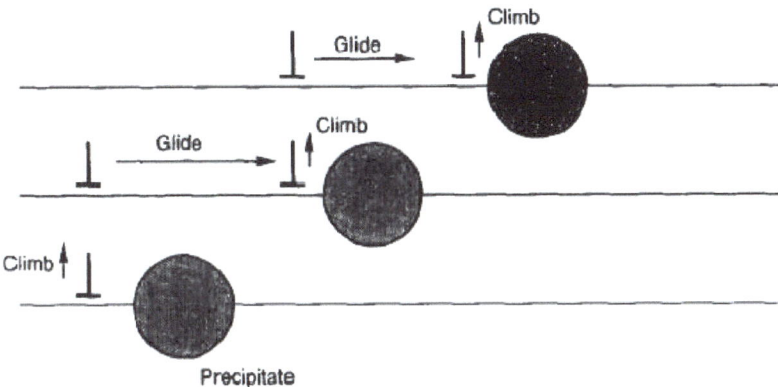

Fig. 41. How the climb-glide sequence leads to creep [66].

This explains the progressive, continuous, nature of creep, and the role of diffusion, with diffusion coefficient:

$$D = D_0 \exp^{-\frac{Q}{RT}} \quad \text{----------------(7)}$$

2. 6. 1. 2. Diffusion creep (giving linear-viscous creep):

As the stress is reduced, the rate of power-law creep falls quickly, but creep does not stop; instead, an alternative mechanism takes over. As Figure 2.42 shows, a polycrystal can extend in response to the applied stress (σ) by grain elongation; here, σ acts again as a mechanical driving force but, this time atoms diffuse from one set of the grain faces to the other, and dislocations are not involved. At high T/Tm, this diffusion takes place through the crystal itself, that is, by bulk diffusion. The rate of creep is then obviously proportional to the diffusion coefficient D, and to the stress σ (because σ drives diffusion in the same way that dc/dx does in Fick's Law); and the creep rate varies as $1/d^2$ where d is the grain size (because when d gets larger, atoms have to diffuse further). Assembling these facts leads to the constitutive equation:

$$\dot{\varepsilon} = C \frac{D\sigma}{d^2} = \frac{C'.\sigma.\exp^{-Q/TR}}{d^2} \quad \text{----------------(8)}$$

where C and C' = CDo are constants. At lower T/Tm, when bulk diffusion is slow, grain-boundary diffusion takes over, but the creep rate is still proportional to σ. In order that holes do not open up between the grains, grain-boundary sliding is required as an accessory to this process [66].

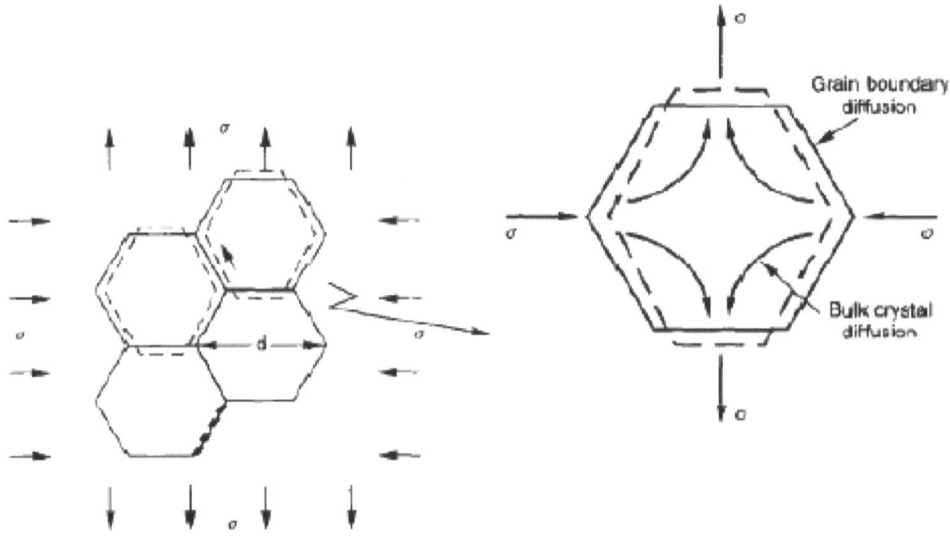

Fig. 2.42: How creep takes place by diffusion [66].

2. 7. Characteristics of 9-12% cr creep resistant steels:

The creep resistance of a CrMo steel is based on the formation of stable precipitations such as alloy carbides in a ferritic, bainitic and/or martensitic microstructure in the normalised condition. Due to a subsequent tempering treatment, a stable microstructure with precipitations is generated that remain stable at the service temperature for which the steel has been developed. The precipitations formed will block the grain-boundaries and prevent sliding of the slip-planes to give the desired creep resistance properties. They should therefore have the correct shape, be present in the right amount and be evenly distributed to obtain a homogeneous structure with homogeneous properties. Depending on the alloy level and the heat treatment(s), specific types of precipitations will be formed in a specific amount. The governing parameters for the heat-treatment are temperature and time [47]. The variety of precipitations that can be expected and that are mainly used in the design of classic and modern creep resistant CrMo steels are listed in Table 2.8.

Table 2.8: Precipitations that can be found in creep resistant CrMo steels[47].

Precipitations and possible phases in CrMo steels		
Graphite		
Epsilon	= $Fe_{2.4}C$	
Cementite	= Fe_3C	
Chi	= Fe_2C	
M_2X	M_6C	$M_{23}C_6$
M_7C_3	Laves	
M_5C_2	Z-phase	
Mo_2C	Cr_3C	
NbC	NbN	VN

2. 7. 1. Long-term testing:

One of the most demanding tasks for the validation of new high temperature steels for use in pressure vessel design is the development of a comprehensive database of long-term creep test results on industrial products. For approval by the ASME new steels need tests with rupture times of at least 10,000 hours to establish allowable stresses. For approval in Europe rupture times of at least 30,000 hours for five melts of a new steel are needed to make a valid extrapolation to the 100,000 hour mean rupture strength to be used for design. This requirement is the main reason that the lead time from laboratory to application for a new high temperature steel is approximately 10 years as demonstrated by the examples above [49].

Table 2.9 : Statistics of ECCC databases for steels P92, E911 and P91 [39].

ECCC database statistics
Grade 92 - 2004

Test temperature °C	Number of test results in time interval					
	<10,000h	10,000h-20,000h	20,000h-30,000h	30,000h-50,000h	50,000h-70,000h	>70,000h
550	92	4(2)	4	(1)	1(2)	
575	9(1)	2(1)	1	2	(2)	(1)
600	196(3)	30(1)	17(1)	16(2)	8(4)	3(1)
625	33	9	2(1)	1(1)	1	
650	211(1)	26	15	15(2)	4	2
675	18					
700	122	5(1)	1			
750	17					

Parantheses denote unbroken tests   Total testing hours: 7,036,945

E911-2004

Test temperature °C	Number of test results in time interval					
	<10,000h	10,000h-20,000h	20,000h-30,000h	30,000h-50,000h	50,000h-70,000h	>70,000h
550	13	1	2(1)		1(2)	
575	25(1)	4(1)	3	2	2	
600	86(1)	21	8	4(2)		(1)
625	49(3)	8	2	1	1	
650	75(8)	14	3(1)	2(1)	(1)	
675	6					
700	2					

Parantheses denote unbroken tests   Total testing hours: 3,148,115

P91-2005

Test temperature °C	Number of test results in time interval					
	<10,000h	10,000h-20,000h	20,000h-30,000h	30,000h-50,000h	50,000h-70,000h	>70,000h
500	71(6)	6(3)	2(6)		1(3)	(1)
525	3(3)		(1)		(1)	(2)
550	262(11)	31(9)	21(7)	12(2)	6(2)	5(1)
575	58(1)	10(1)				(1)
600	524(7)	99(11)	43(7)	29(4)	9(5)	2(8)
625	57(1)	2(1)				
650	358	40(4)	16(5)	6	6	(2)
675	6					
700	40	4				

Parantheses denote unbroken tests   Total testing hours: 17,852,130

In Europe long-term creep tests are carried out and coordinated by the European Creep Collaborative Committee ECCC, where testing efforts are shared between a number of countries, and long-term strength values for European codes and standards are established according to commonly agreed extrapolation methods [48].

For steels P91, E911 and P92 the ECCC have made worldwide data collations resulting in extensive databases, see Table 2.9. In 2005 new evaluations of the data have been made resulting in ECCC data sheets with validated rupture strength values for the steels up to 100,000 hours. Compared to previous evaluations made by the ECCC 8-10 years ago the new evaluations show a slight reduction of 8-10% in the 100,000 hour strength values for all three steels. This is due to additional data from further suppliers leading to higher scatter and not least to substantial additional long-term data for the steels in the range of more than 30,000 hours.

The ECCC evaluations have been submitted to the ECISS (European Committee for Iron and Steel Standardization) for inclusion in the relevant steel standards. The comprehensive data from long-term testing has confirmed excellent creep stability and sufficient creep ductility of all three steels up to 600°C [39]. On the other hand ASME consider the obtained results through the new 2009 addenda for the ASME Boiler and Pressure Vessel code 2007; and all data related to allowable stress, heat treatment, welding, construction, …etc; have been added [49].

2. 7. 2. Thermal fatigue resistance:

Fatigue due to cyclic loading can occur at local stress concentrations and thermal stresses due to temperature differentials. The latter is of particular relevance to thick walled vessels which may show rapid changes of temperature due to start-up and shut down cycles. Hot water or steam entering a cold, thick walled vessel can lead to steep temperature gradients through the wall thickness. Expansion of the inner regions of the wall, close to the bore, may be constrained by the outer, cooler material. This imposes compressive stresses near the bore which may be sufficient to cause local yielding. As the temperature gradients equalize, and the outer regions of the wall expand, the material at the bore may then go into tension. With continued operation at temperature the residual tensile stress relaxes until, on cooling, the process is reversed. Principal factors in limiting the thermal fatigue life of plant components are the material properties (thermal conductivity, coefficient of expansion, mechanical properties), plant start-up and shut-down rates and the presence of stress concentrating factors, such as tube penetrations into a header. Ferritic steels generally have higher thermal conductivities and lower thermal expansion coefficients than austenitic steels and are thus less prone to suffer from thermal fatigue (Figures 2.43 & 2.44) [50].

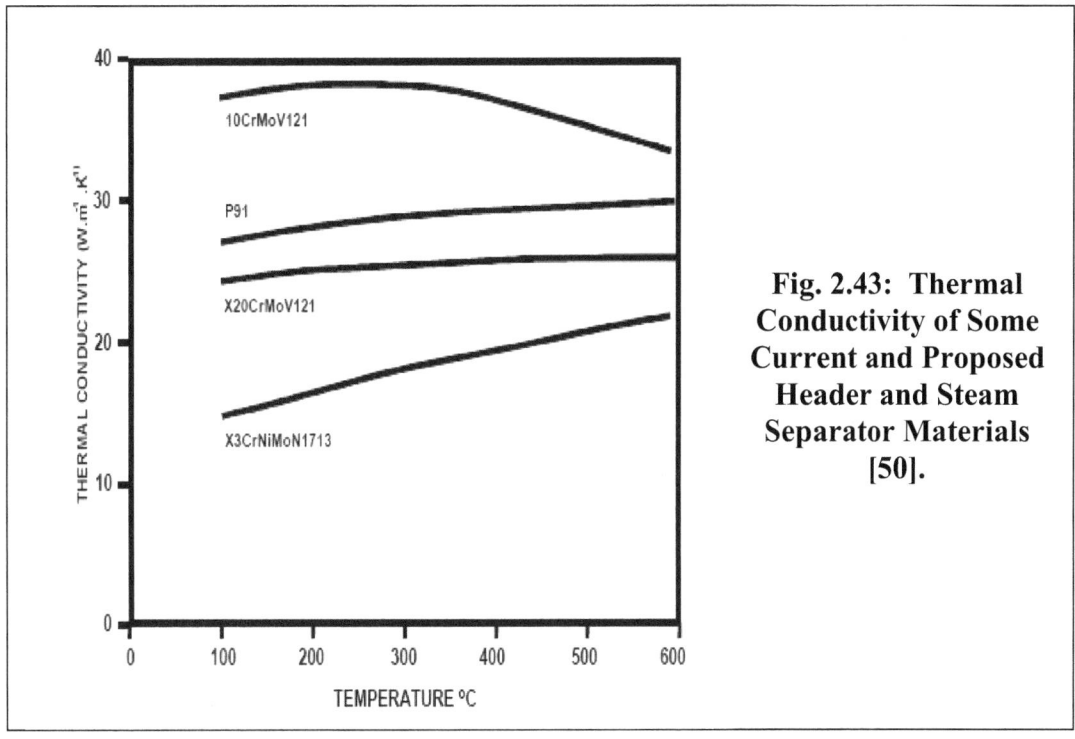

Fig. 2.43: Thermal Conductivity of Some Current and Proposed Header and Steam Separator Materials [50].

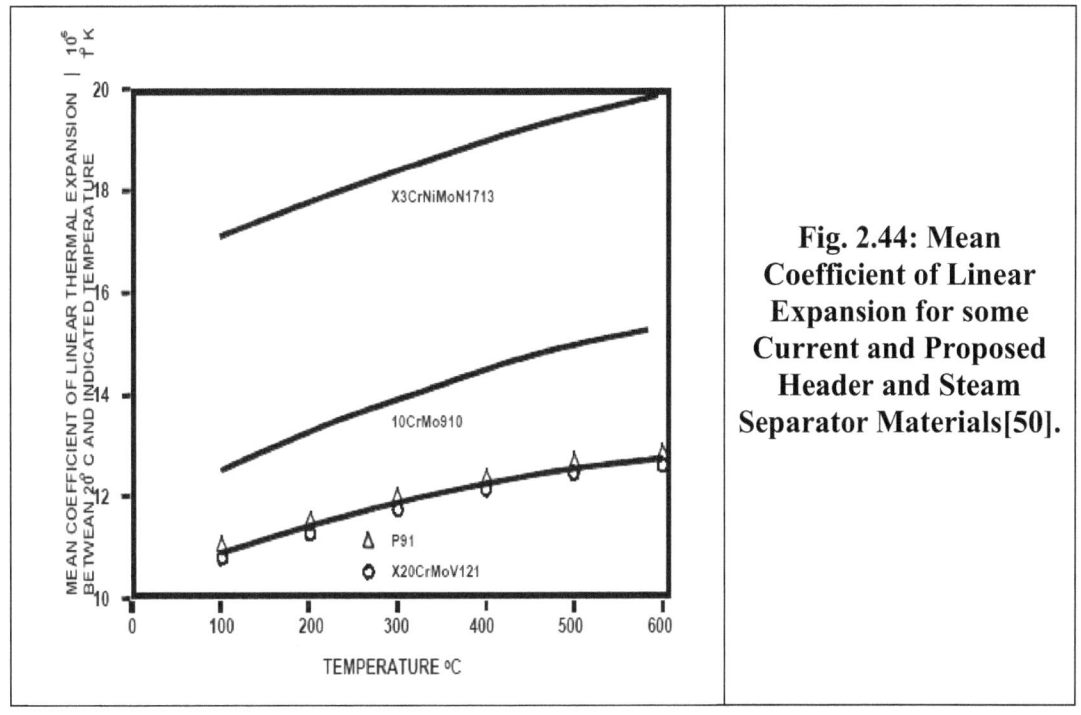

Fig. 2.44: Mean Coefficient of Linear Expansion for some Current and Proposed Header and Steam Separator Materials[50].

2. 7. 3. Steam oxidation resistance:

Almost without exception, high temperature materials rely on the selective oxidation of one or more alloy constituents to form a protective oxide scale. Two conditions need to be fulfilled; firstly, there must be a sufficiently high concentration of the selectively oxidizing elements in the matrix and secondly, the diffusion rate of these elements must be fast enough to ensure that they replenish the matrix below the growing scale, thus ensuring long-term protection.

In the high Cr steels, clearly Cr is the most important constituent with regard to oxidation resistance. In the development of the modified 9Cr1Mo steels, the emphasis was on improvements in the stress rupture strength. Long-term creep tests carried out at temperatures up to 650°C had shown that the oxidation resistance in air was excellent due to the formation of tightly adherent, protective oxide scales. The protective scales formed in air on 9% Cr steels were identified as consisting of $(Fe,Cr)_2O_3$ and $(Fe,Cr,Mn)_3O_4$. However, in steam-containing atmospheres, the scales formed at 600 and 650°C were thick and consisted of an external Fe3O4 scale and an internal two-phase scale of Fe_3O_4 and $(Fe,Cr,Mn)_3O_4$ [51]. Below the oxide scale, internal oxidation of Cr to form $(Fe,Cr,Mn)_3O_4$ occurred at the martensite lath boundaries.

Figure 2.45 shows typical oxide scales formed on the 9% Cr steels in steam-containing atmospheres at 650°C and Figure 2.46 summarizes the mass changes that occur.

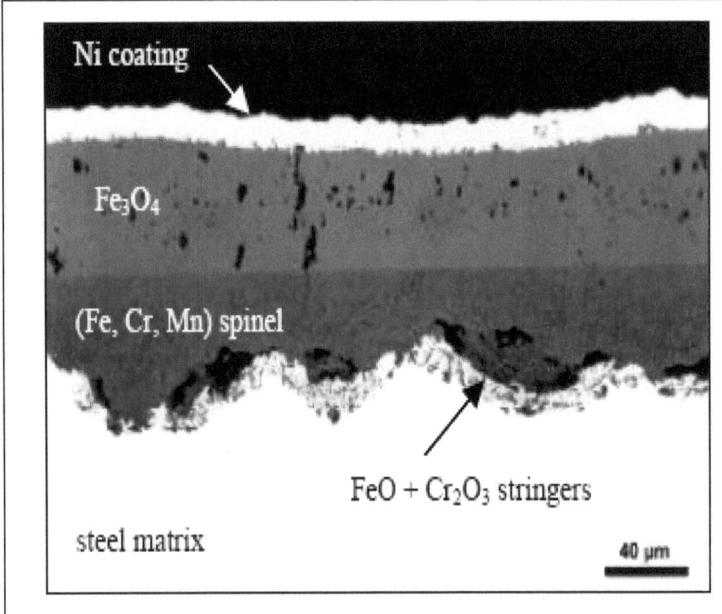

Fig. 2.45.: Scale microstructure of P91 exposed for 1 000 h at 650°C in Ar-50% H₂O.

There are several concerns about the high oxidation rates seen in steam. Firstly, the loss in load-bearing cross-section due to the oxide scale formation and the internally oxidized zone will lead to stress increases and therefore a reduction in service lives. The reduction in life is, of course, dependent on the initial wall thickness of the components under consideration. Calculations reported by Quadakkers and Ennis 1998 [52]; have shown that the life reduction at 600°C in steam will be significant for components with wall thicknesses below about 6 mm.

A second, and more difficult to quantify, effect is the spelling of the oxide. The presence of spelled oxide particles in the steam entering the turbine can cause erosion problems and local blockages. Furthermore thick oxide scales on heated tubes can lead to a decrease in the heat transfer across the tube walls, resulting in overheating and subsequent creep failure of tubes.

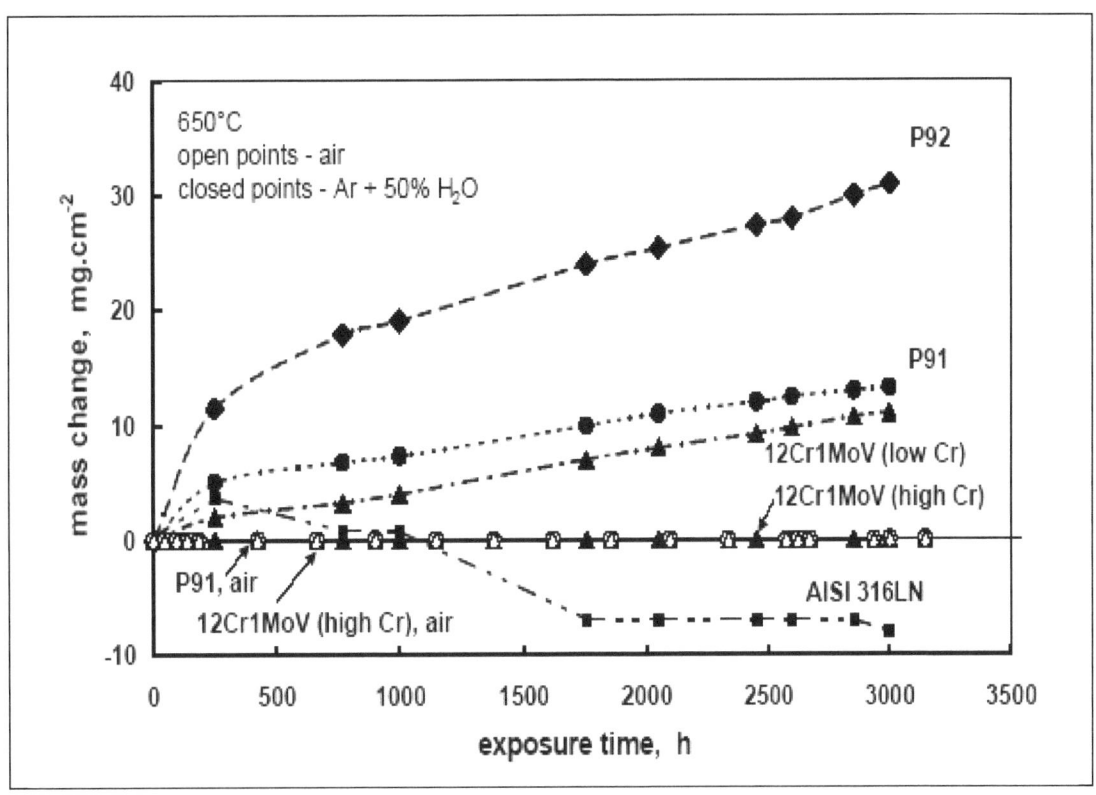

Fig. 2.46 : Mass change data for Cr steels exposed in Ar-50% H2O and in air at 650 °C.

In order to achieve acceptable oxidation resistance at 600 and 650°C, according to Ehlers et al 2001, a Cr content of at least 11% is required, to enable the formation of a protective spinal scale [51]. The oxidation resistance may be enhanced further by the addition of selectively oxdisable elements, the main contenders being Si and Mn. Si is, however, a ferrite former, and when added to the steel in amounts that are clearly beneficial for the steam oxidation resistance can lead to toughness and fabrication difficulties. There are some additional factors that need to be considered in assessing the steam oxidation resistance of the high Cr steels. The effect of steam test parameters, such as flow rate and pressure, are not sufficiently well established and there is a need for further investigations to ensure that laboratory data reflects sufficiently well the expected behavior in a power plant. The steam oxidation resistance of the high Cr steels is enhanced by [53]:

- Cr content of at least 11%;
- the addition of oxygen active elements, such as Si and Mn;
- increasing the diffusion of Cr to the surface by suitable bulk alloying additions or by surface deformation treatments

2. 8. Heat treatment and microstructure of 9-12% cr steels :

The final heat treatment for both T/P91, T/P92 and T/P911 consists of normalizing and tempering. An austenitising temperature of about 1060°C is adopted for hardening. A fully martensitic structure is obtained upon cooling to room temperature over a wide range of cooling rates. The microstructure and its stability to the aging effect had been discussed in detail under paragraph 2.3.

2.8.1. Continuous cooling diagrams and transformation behavior:

The Continuous Cooling Transformation (CCT) diagram of T/P91 and T/P911, shown in Figure 2.47, is essentially the same as that of T/P92. The M_s and A_{C1b} temperatures as well as martensite hardness are more or less the same for the three steels. And also the location of the ferrite carbide transformation nose is comparable to that of T/P91 [13].

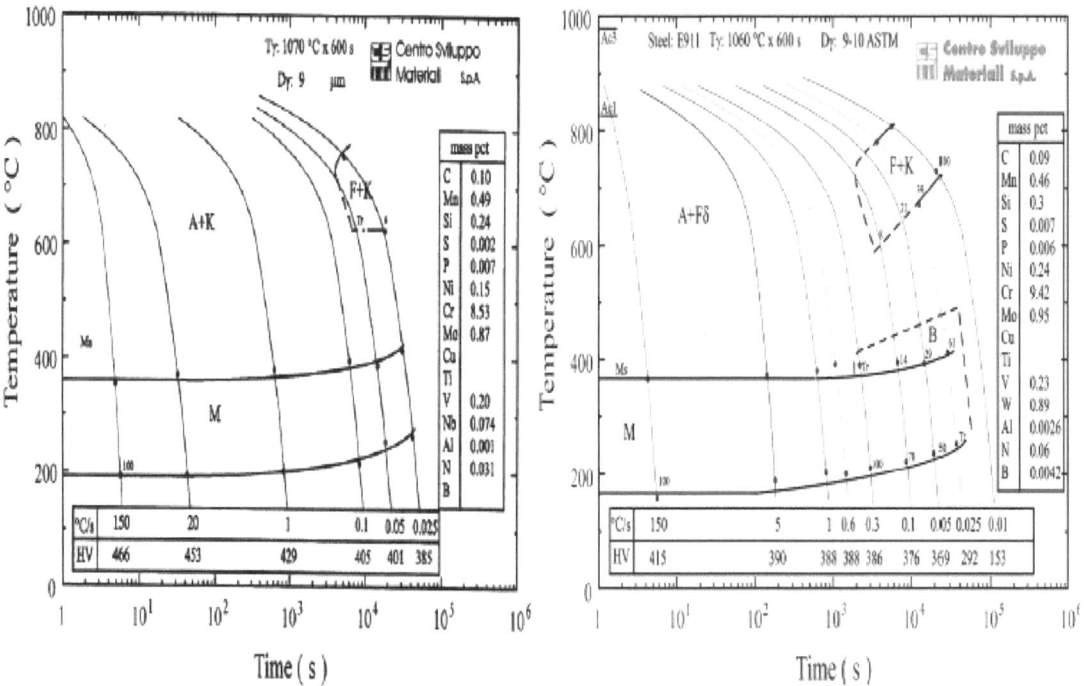

Fig. 2.47. CCT diagram: (a) Grade 91. (b) Grade 911 [34].

In both steels a martensitic microstructure is achieved by cooling from normalizing temperature in a very wide range rates >0.2 °C/s. The microstructures are characterised by hardness values between 405-460 HV, depending on component thickness. In order to comply to ASTM and obtain a good compromise among creep resistance, toughness and ductility (i.e. cold formability), tubes and pipes were finished in normalized and tempered conditions. The normalizing treatment produces the desired martensitic microstructure and provides a good carbide and nitride solubilisation into the matrix. With the subsequent tempering the martensite is recovered and the precipitation of $M_{23}C_6$ carbides and fine MX carbo-nitrides is promoted, according to the following precipitation sequence [34]:

$$M_3C \rightarrow [M_{23}C_6 + M_2X] \rightarrow [M_{23}C_6 + MX] \quad \text{------------ (9)}$$

The microstructures of Grade 91 and Grade 911 pipes after normalizing and tempering are respectively shown in Figure 48-a and 48-b. It can be observed that a small amount of δ-Ferrite is present in 911 steel (<3%). The quantity of δ-Ferrite which forms as a function of temperature was measured with laboratory heat treatments: the normalization at 1060°C gives the minimum amount of ferrite (Figure 2.49).

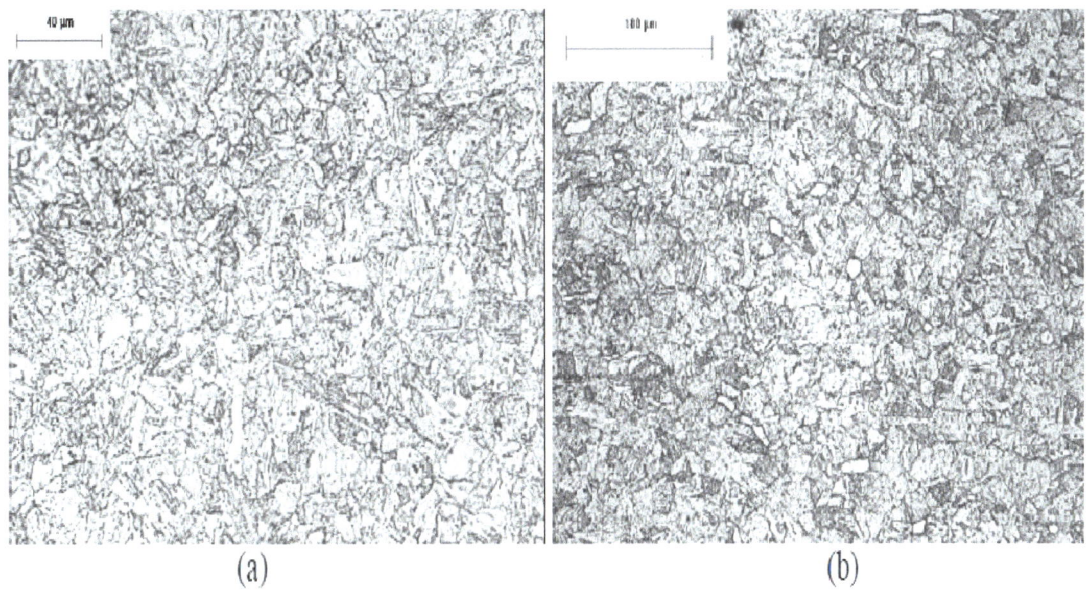

Fig. 2.48 – Samples of (a) Grade 91 after 1070°C+780°C and (b) Grade 911 after 1060°C+780°C [34].

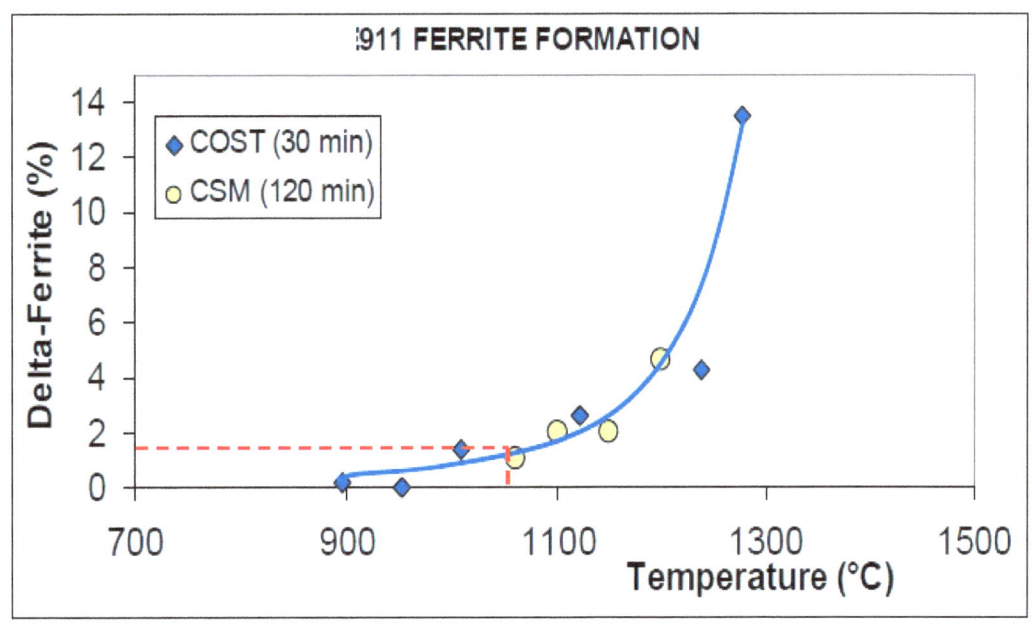

Fig. 2.49.: Delta-ferrite fraction in E911 steel as function of temperature. COST experimentation (30 min holding time in furnace); and CSM experimentation (120 min holding time) [34].

2. 9. Weldability and welding of 9-12% cr steels :

The weldability of chromium-molybdenum steels is very similar to that of quenched, tempered and hardenable low alloy steels. The major problem in the HAZ is cracking in the hardened coarse-grained region, as well as HAZ softening between Ac_1 and Ac_3. Reheat cracking during PWHT and long-term exposure in elevated temperature service conditions also can cause severe problems. However, because these steels are a high-strength alloys that normally transforms completely to martensite during air cooling, the specification of preheat and post weld heat treating (PWHT) conditions is an important practical matter[54].

In the weld metal and part of the heat affected zone (HAZ), where the α-γ-α\ transformation occurred, the proportion of martensite is always considerable; therefore, during welding, a certain risk of cracking may need to be considered. As a consequence of these arguments, welding engineers may need to have a general knowledge of the most important microstructural transformations and the related changes in strength. The quantity of martensite transformed from austenite depends on the temperature under cooling below the MS. This transformation is almost independent of the composition, and takes place in all steel grades as shown in Figure 2.50. Initially, this relationship has a linear characteristic, and it is suggested that this may be estimated by a straight line by which the transformation is completed at MS – 126 °C. In this case the volume fraction of martensite is above 90 %, where the transformation slows down and finishes at approximately Mf = (MS – 190) ±10 °C [56].

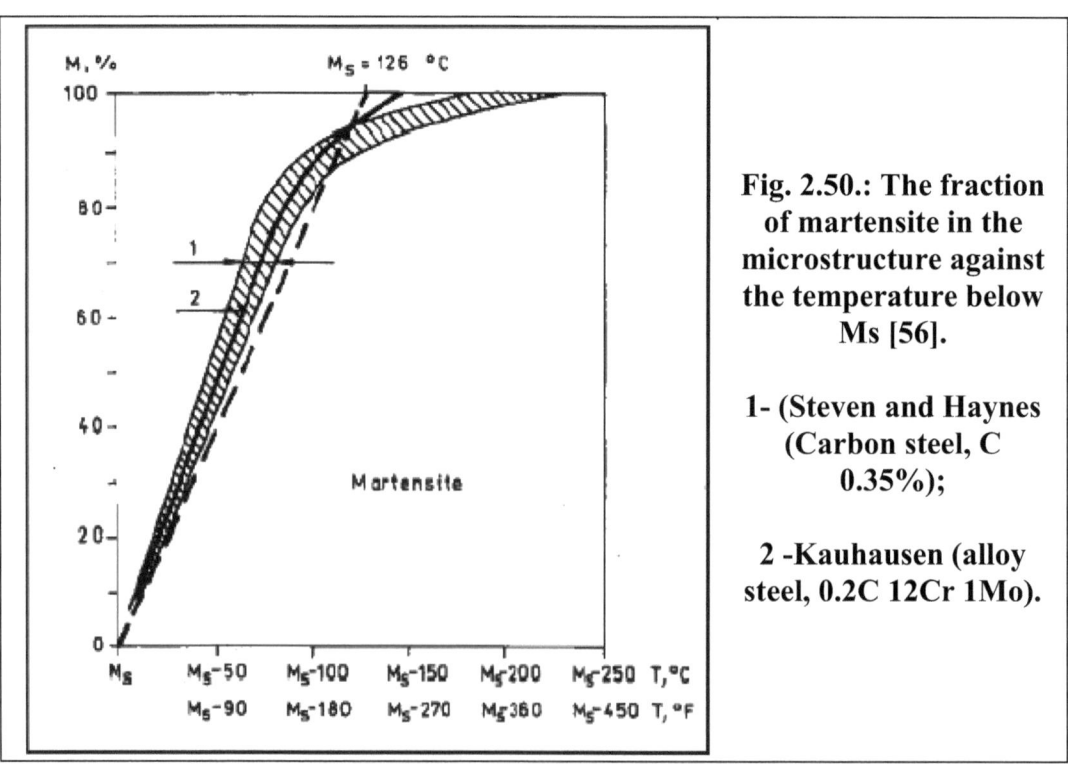

Fig. 2.50.: The fraction of martensite in the microstructure against the temperature below Ms [56].

1- (Steven and Haynes (Carbon steel, C 0.35%);

2 -Kauhausen (alloy steel, 0.2C 12Cr 1Mo).

According to Figure 2.50; when welding is carried out at the preheat temperature MS-25 °C, approximately 20 % of martensite can be found in the microstructure of previously austenitized steel, but at MS - 75 °C, the proportion of martensite reaches 60

%. This may mean a 50 °C difference in preheat temperature increases the martensite concentration up to three times [56].

One aspect for the handling of this issue is the handling of weldments between completion of welding and PWHT. Weldments that are cooled to room temperature prior to PWHT will transform more completely to martensite than weldments that are maintained at or above minimum preheat temperature prior to PWHT. Consequently, weldments that are cooled to room temperature are less likely to contain un-tempered martensite after PWHT. However, maintaining preheat temperature prior to PWHT is essential for minimizing the probability of hydrogen cracking in weld heat affected zones. In addition, it makes more efficient use of energy resources, and it may be desirable for practical reasons, depending on the welded component size and the available facilities for material handling and heat treating [55].

Hot cracking is another problem with these steels that can occur with the combination of high sulfur, carbon, and nickel together with high restraint. Finally, stress relief cracking is of concern when welding quenched and tempered steels and heat resistant steels that contain significant levels of carbide formers such as chromium and vanadium [54].

2. 9 .1. Avoidance of hydrogen induced cracking:

Generally for high alloyed steels, obtaining a welded joint free of cracks requires preheating the joint before the welding process starts. The main aim of this preheating is to avoid the cold cracking caused by the hydrogen induced in the weld and the fast cooling of the joint. The upper limit of preheating has not to be exceeded in order to avoid the grain growth of the welded structure. Control of the temperature of a weld can prevent problems related directly to welding, or can reduce the severity of some effects. In any welded joint, the following problems may arise:

- Hydrogen-induced cold cracking may occur in the HAZ or possibly the weld metal, of partially completed weld.
- Residual stresses may locally exceed the yield strength of the materials.
- Restraint or rigidity of the overall structure may contribute to the generation of reaction stresses transverse to the direction of welding.
- Heat of welding may produce unacceptably low toughness or mechanical properties in the weld area compared to the base plate.

Preheating is specified primarily to prevent hydrogen-induced cracking. Preheating reduces the cooling rate of the HAZ of each weld bead. Slower cooling rates reduce hardness, hydrogen content in the vicinity of the weld and stresses. Preheating temperature is maintained throughout the welding cycle and is known as the interpass temperature between weld beads. Hydrogen cracking can occur if the interpass temperature becomes too low. An upper limit on the interpass temperature also must be maintained because if it becomes great, then excessive grain growth in the HAZ may result, with a corresponding reduction in mechanical properties [54].

2. 9. 1. 1. Calculation of preheating temperature:

In order to prevent welding imperfections of the welded joints of these steels an optimum method for determining the preheating temperature has to be applied. During the last decade, many investigations have been carried out on this topic and a number of results have been published. One of these methods was suggested by L. Beres [56]; that based on the metallurgical knowledge and particularly the chemical composition of the steel to be welded. L. Beres and his co-workers suggested also that the preheat temperature for the martensitic welding of creep resistant martensitic steels should be equal to the interpass temperature. For 0.2 % carbon content steels, preheating temperature is calculated by Equation (10).

$$TP0.2 = (MS - 60) \pm 10 \text{ °C} \quad \text{---------------- (10)}$$

While for 0.1 % carbon content steels in the same group, Equation (11) is used.

$$TP0.1 = (MS - 90) \pm 10 \text{ °C} \quad \text{---------------- (11)}$$

Where MS can be computed using Equation (12).

$$M_S = 454 - 210C + \frac{4.2}{C} - 27Ni - 7.8Mn - 21Cu - 9.5(Cr + Mo + V + W + 1.5Si) \quad \text{-------------------- (12)}$$

2. 9. 2. Heat treatments for CrMo steels and welded joints:

The heat treatments for the base materials are reasonably complex but are required to obtain the optimal mechanical properties. Depending on the alloy content a Normalizing, Tempering and Annealing treatment at various temperatures for several hours with a controlled cooling rate have to be executed according strict procedures. The same is valid for the weldmetal, with increasing alloy content the Post Weld Heat Treatment (PWHT) for welded joints gets more complicated as illustrated in Figure 2.51.

When in subsequent PWHT, Intermediate Stress Relieving (ISR) or in service, the ultimate heat treatment temperature of the base material is exceeded too much and too long, the precipitations can dissolve again which causes reduction of the mechanical properties of the base material. This implies that for example for this reason the maximum temperature of 760°C for P91 in Figure 2.51, shall not be exceeded [47].

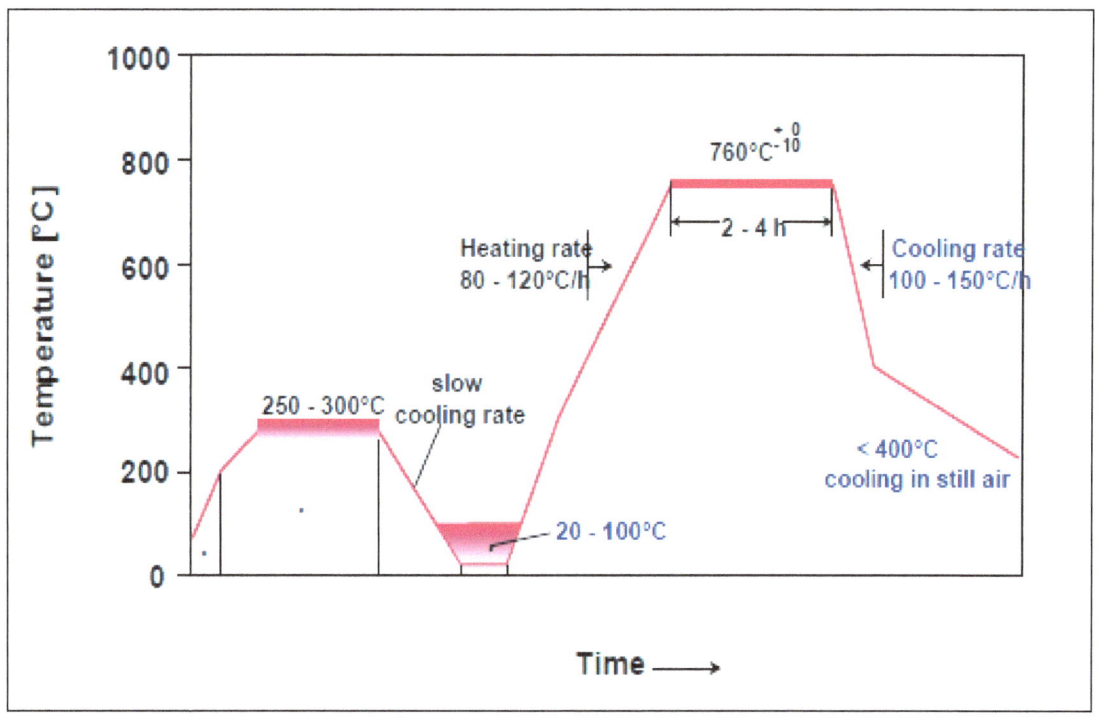

Fig. 2.51: Temperature cycle and heat control during welding and PWHT of martensitic steel P91, E911 and P92[47].

2. 9. 2. 1. Intermediate heat treatment:

In welding thick sections, which are prone to hydrogen-induced cracking, an intermediate heat treatment sometimes is used to reduce peak hydrogen concentration. The temperature of the weld zone is increased to 625 °C for approximately 1 h in a typical heat treatment cycle. An extreme case involves an intermediate heat treatment after each weld bead [54].

2. 9. 2. 2. Temper embrittlement:

When CrMo base material and the weld metal is exposed to a temperature range of 400-500°C for a very long time there is a risk of Temper Embrittlement. This is type of embrittlement is caused by trace elements as P, Sb, Sn and As that migrate to the grain boundaries and can reduce the ductility in both base material and weldmetal. To which extend this phenomena will occur depends merely on temperature and time. To establish the sensitivity of a material to temper embrittlement, a Step Cooling (STC) heat treatment is carried out in the range of 593-316°C for duration of 240 hours. The difference in transition temperature (impact properties) from before and after the heat treatment is a measure for the sensitiveness to temper embrittlement. A maximum allowable shift in transition temperature after step cooling can be specified as a requirement for base material and weldmetal. In order to reduce the risk of temper embrittlement, the responsible trace elements need to be restricted. Bruscato and Watanabe have developed formulas to express the tendency of temper embrittlement :

Bruscato: $X = (10P + 5Sb + 4Sn + As) / 100$ element in wt% and result in ppm   ---------------- (14)

Watanabe: $J = (Mn + Si) \times (P + Sn) \times 10^4$ elements in wt%   ---------------- (15)

The formula of Watanabe is only valid for the base material and is usually restricted to a value of J < 160 but also requirements for J < 120 or 80 are being specified by the industry today.

The Bruscato formula, also referred to as the X-factor, is valid for both weld metal and base material. For weldmetal the specifications are becoming more and more stringent with increasing wall thickness and desire for additional assurance of the mechanical properties. Initially the required value of the X-factor was X < 15 but present specifications already ask for X < 10. An additional requirement for the Mn and Si content can be set according: Mn + Si < 1.1% Specifically for SAW where the trace elements can be picked up from both wire and flux, the combination should be tested to comply with the requirements. This means one source for both wire and flux would be recommended [47].

2. 9. 3. Welding and welding consumables for CrMo steels:

In general creep resistant CrMo-steels are welded with matching consumables in order to have a homogeneous welded joint with about equal mechanical properties. Matching compositions also have the same coefficient of thermal expansion, which prevents or at least reduces the risk of thermal fatigue in service. In this respect, the heat affected zone (HAZ) is a vulnerable area. In principle all arc welding processes can be applied as SMAW. GTAW, GMAW, SAW and FCAW. For the manual processes it is important to take sufficient measures to protect the welders from heat then it is of utmost importance that the preheat as well as the interpass temperatures are respected and not reduced to accommodate the welders, also while tacking. With the gas-shielded processes it is vital to assure proper shielding of the weld. Due to the high preheat, the gas-shield can be distorted and provide less protection as required. Special nozzles and gas cups are available to reduce the problem [47].

2. 9. 4. Characteristic of welded joint :

Due to the heat input during welding, microstructural changes take place in a small area (depending on the welding process, welding geometry and material type) besides the fusion line. If the A_1 temperature is exceeded in that area, phase transformations lead to changes in the microstructure of ferritic steels, which can clearly be seen using optical microscope, that are called heat affected zone (HAZ). Thus, the HAZ does not represent the optimal microstructure and precipitation characteristics that can be found in the unaffected base material (BM) needed for optimized creep resistance. Furthermore, according to standards and manufacturers specifications, the unaffected base material has to be subjected to a specific heat treatment for maximum creep strength.

Weldments can be considered as a structure with 5 materials (base material, weld material, 3 heat affected zones) with different behaviors (Figure 2.52). The behavior of a structure is geometry dependent, since these different material properties cause a complex stress and strain distribution. Creep and stress redistribution is influenced by this situation [57].

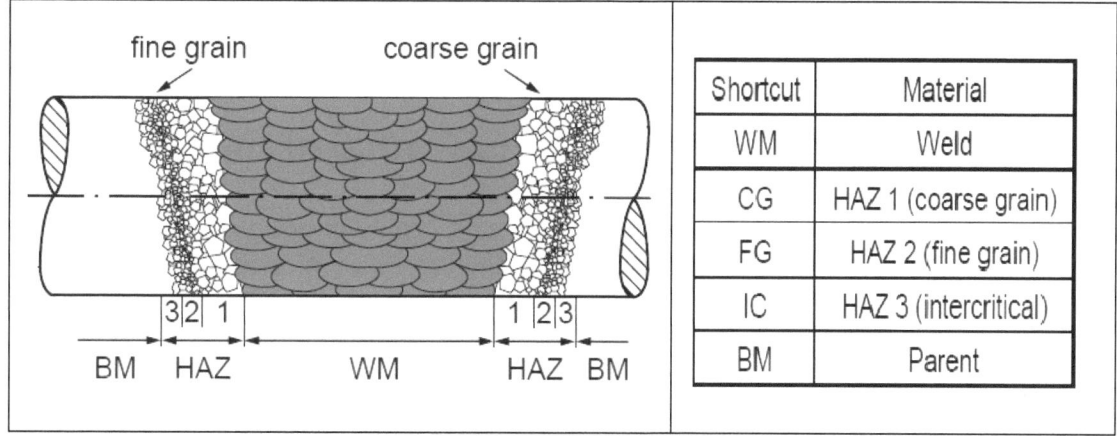

Fig. 2.52. Weldment different regions[57].

2. 9. 4. 1. Parent metal (PM) :

The parent metal (PM) or the base metal (BM) is the metal to be welded. The PM region, except near the fusion boundary is unaffected by the welding process, and therefore it is not metallurgically affected by welding. However, due to the welding process the PM is likely to be in a state of residual transverse and longitudinal shrinkage stress [54], of magnitudes that depend on the degree of restraint imposed on the weld.

2. 9. 4. 2. Heat affected zone (HAZ):

The HAZ is a part of PM that has not been melted by the welding process but in which the microstructure and mechanical properties have been altered [54]. The heat affected zone (HAZ) is the area that has been microstructurally affected by the heat as a result of the thermal cycle during welding. Because of varying thermal conditions as a function of distance from the fusion line, the HAZ is composed of a structural gradient that can be described in term of the GCHAZ, the GRHAZ and ICHAZ, as mentioned in 2.9.3.

Because grain growth is a function of temperature and the peak thermal cycle temperature decreases sharply with distance from the fusion zone, the maximum austenite grain size always occurs near the weld fusion boundary and decreases with distance from the boundary. The major factor that determines the maximum grain size is the residence time at elevated temperatures where rapid grain growth can occur.

2. 9. 4. 2. 1. Grain-refined heat affected zone (GRHAZ):

The GRHAZ is defined as the region where the transformation of carbide and ferrite to austenite is completed, but there is negligible or little grain growth. In lower C, low alloy steels, this austenite decomposes into small pearlite colonies and ferrite grains[54]. In high hardenability steels like P91, martensite can form despite the small grain size. The GRHAZ is also referred to as the fine-grained (FG) HAZ.

2. 9. 4. 2. 2. Grain-coarsened heat affected zone (GCHAZ):

The GCHAZ is defined as the region of coarse grains adjacent to the WM. The microstructure of this region results from the large austenite grains that form due to the temperature of this region being well above the Ac_3 temperature [58].

2. 9. 4. 2. 3. Intercretical heat affected zone (ICHAZ):

The intercritical HAZ is defined as the region directly adjacent to the PM running parallel to the fusion line where the peak temperature is sufficient to cause partial austenitisation. When a weldment fails in this region under creep conditions it is referred to as a Type IV failure [58].

2. 9. 4. 3. Weld metal (WM) :

The WM zone comprises a mixture of the filler metal and base metal, which is completely melted and relatively homogenous. At the outer boundary of the WM region is the weld interface zone. This is defined as the boundary between the un-melted base metal on one side (HAZ) and the completely fused weld metal on the other side [54]. Figure 52 shows a schematic representation of the different zones in a typical butt weld and the corresponding temperature profile. In this figure the GCHAZ is referred to as the grain growth zone, the GRHAZ is the re-crystallised zone, and the ICHAZ is the partially transformed zone.

2. 9. 4. 4. Dissimilar welding of 9-12% Cr steels [72]:

Dissimilar welds involving Creep Strength-Enhanced Ferritic Alloys (CSEF's), austenitic stainless steels and other alloys are performed on a routine basis. Welds between low alloy ferritic steels and the CSEF's utilize either Grade 91- type filler (B9), materials matching the lower alloy type base metals, or nickel based alloys. Dissimilar welds between CSEF's and other alloys are an area where research is ongoing. EPRI report TR 1009758, *Evaluation of Filler Materials for Transition Weld Joints between Grade 91 to Grade 22 Components*, found that using 2-1/4Cr 1Mo (-B3) weld filler metals provided best results where similar thickness were involved whereas nickel base (NiCrFe-3) alloys were actually too strong and may cause carbon migration. Use of –B9 (P91) weld metal was acceptable, but there were very strong and require tempering at high temperatures. Factors that can occur in dissimilar metal welds include: carbon diffusion (migration), stresses from thermal expansion, sensitization of base materials, notch effects from differences in strengths, etc.

Carbon diffusion can occur during the tempering heat treatment due to the difference in chromium content of the materials. Carbon will migrate from the lower chromium material to the higher chromium containing material. For example: When P22 (E901X-B3/ER90S-B3/EB3) filler metal is used, the decarburized zone will be in the P22 weld metal next to the P91 base metal with the carburized zone located in the P91 HAZ. If P91 weld metal is used, the carbon depleted zone will be located in the coarse grained P22 base metal HAZ and the carburized zone in the P91 weld metal. The extent of the decarburized zone depends on the tempering temperature and time at temperature. Using a nickel-base welding consumable typically minimizes this condition although higher chromium containing nickel base alloys are still affected.

Transitions or dissimilar welds between CSEF's and austenitic stainless steels normally use nickel-base weld material. The weld can be completed by buttering the CSEF with a nickel-base filler metal, followed by a normal CSEF PWHT. The weld between the PWHT'd nickel butter and the austenitic stainless steel then can be completed using nickel-base filler metal without PWHT. The reason for going to the expense of performing a butter followed by a PWHT then completing the weld is to keep the stainless steel side of the weld from sensitizing during the PWHT operation. Nickel base weld metals typically used include: ENiCrFe-2 & -3, ERNiCr-3 and EPRI P87 alloy. Research indicates that P-87 will clearly offer better results.

To enhance the viability of using a nickel base alloy, EPRI evaluated and developed EPRI P87 filler metal. EPRI P87 filler metal is a nickel base filler metal that only contains about 9% chromium so that along with other benefits, carbon migration is significantly reduced. This alloy has the potential to eliminate PWHT for field applications when weld grooves are buttered with EPRI P87 and receive PWHT in the shop. Further, this approach provides a mechanism to perform complete normalize and temper heat treatment to eliminate the Type IV region.

Another option is to use a short transition piece. In this method, one weld is made and PWHT'd as a CSEF to nickel-base weld while the weld at the other end of the transition piece becomes a nickel-base to stainless steel weld and does not require PWHT. Alternatively, if sensitization issues are avoided by using a grade of stainless steel that is not subject to sensitization, then a single weld could be made without buttering by using appropriate stainless steel weld metal and P91 post weld heat treatment. Use of transition pieces is usually required where high stresses are present like joining a CSEF to a lower alloy turbine stop valve, where carbon migration or significantly different material strengths are a concern [72].

2. 9. 5. Failur mechanisms of welded joints in the creep range:
2. 9. 5. 1. Ferritic steels – creep rupture:

Dividing up the HAZ into three different zones is a state of the art classification of that area. The coarse grain zone near the fusion line (HAZ1), the area of the outer HAZ (intercritical zone – HAZ3) and an area lying in between with a medium grain size (fine grain zone - HAZ2). These areas show different creep strength and different creep rupture deformation values. In real life of course, each zone has a fluent transition into the next zone.

In general, the coarse grain zone shows relatively high creep strength but rather low creep rupture elongation. The intercritical zone and the transition region into the unaffected base material show a significantly lower creep strength than the unaffected base material. A hardness profile cross-weld usually has a characteristic shape – a maximum hardness in the weld metal and a steep decrease in the HAZ with a hardness trough in the area of the outer HAZ (transition to the uninfluenced base material respectively). The different strength and deformation behavior in the HAZ (stress and strain redistribution) and the related constraint effects as well as the influence of the surrounding material layers, like weld metal (WM) and BM, result in a complex, interacting and multiaxial stress situation in the weld. This particular stress situation in combination with the rather small creep resistance of the HAZ3 results in general in premature creep damage in the intercritical heat affected zone, depending however on the loading and temperature conditions. In case of long term creep loading perpendicular to the weld, a change of fracture location from WM to HAZ3 can be observed and therefore often premature failure of the component [59].

Creep tests on crossweld specimens of different base materials were evaluated [60]. These investigations clearly show the great impact of the base material on the failure behavior of the welded specimens. According to Figure 2.53, unalloyed steels show the smallest difference between the creep rupture strength of the welded joint compared to the parent material's creep strength, whereas bainitic and in particular martensitic steels show a stronger decrease in creep strength after welding. Especially welded joints of martensitic steels, due to their complex microstructure and the related specific heat treatment, are subjected to the greatest decrease in creep strength as a result of the effects described above. Furthermore, a strong influence of the service temperature can be found. It strikes out, that at the maximum service temperatures of the new 9% Cr-steels the factor $R_{m/t/\vartheta}$ welded joint / $R_{m/t/\vartheta}$ base material reaches a value of 0.5, see Figure 2.53.

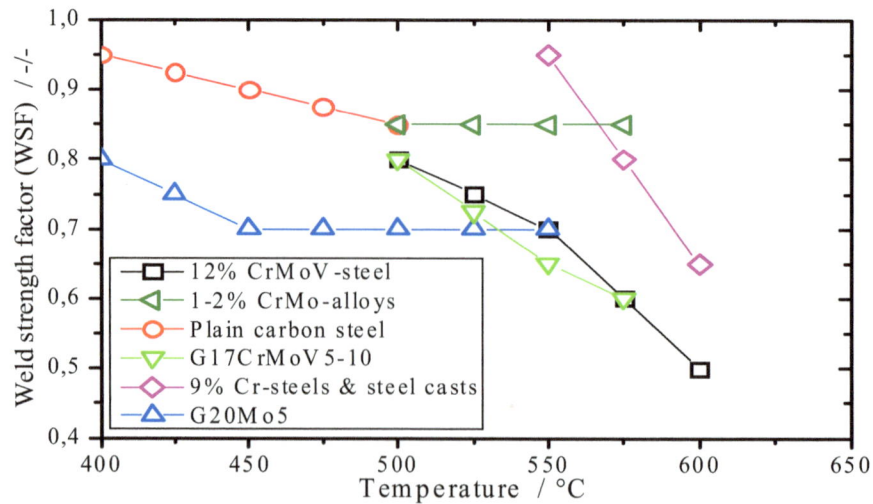

Fig. 2.53: **Weld strength factors for 100.000 h creep rupture strength [60].**

The negative effects on the microstructure and the precipitation characteristics described above can be removed/diminished by performing a complete additional heat treatment (quenching and tempering) of the welded component according to standards or recommended procedures by the steel manufacturer. By doing so, the microstructure of the HAZ is retransformed into the base material's optimized microstructure for maximum creep resistance. If a matching weld metal, with equivalent creep strength comparing to the parent material is used, such components can show a service performance similar to seamless components.

For longitudinally welded pipes there is an alternative technology available, if a complete additional annealing of the component post welding is not possible e. g. due to expense limitations or technical restrictions. The coupling of the submerged arc (SAW) welding process with subsequent continuous austenizing – by inductive heating - of the whole pipe. Tempering can be done in a separate furnace afterwards. Current research on this new technique within the scope of an AVIF research project showed very promising results [61]. Creep tests on P91 (X10CrMoVNb9-1) cross weld specimens, extracted from pipes which were subjected to the heat treatment described above, with running times up to 16.000 h showed, that the change in fracture location and thus premature failure could be avoided. The determined creep rupture data lies well within the scatter band of the base material. There are indications for similar performance of P92 (X10CrWMoVNb9-2) cross-weld specimens (fracture location in the base material at running times of up to 5.000 h and testing temperature of 650 °C) [61].

Challenges to meet of this technique are the relatively short austenizing time and the possibility of a too low temperature (below A_{C3}) on the inside of the pipes due to the inductive heating. Therefore productive measures have to be taken by the pipe manufacturer to guarantee a complete austenizing even on the inside of the pipes. Continuous control and documentation – e. g. by temperature measurements on the pipe's inner surface – seem mandatory for quality assurance.

The evaluation of the microstructural results showed that despite the low austenizing time of only approximately 2 minutes, a positive effect on the microstructure can be found. The results of creep tests with simulated HAZ showed further potential for improvement by further increasing the normalizing temperature, yet limits with regard to height and normalizing time have to be respected. Furthermore it has to be noticed, that a modification of the pipe geometry (wall thickness, pipe diameter) and of the alloy design could require further adjustments.

2. 9. 5. 2. Designation of crack location (types of cracking) - quick over view:

The designations for the characterization of crack locations are illustrated in Figure 2.54. The classification of cracks depending on location and orientation of a crack is given as Type I if the crack location is in the weld metal, Type II if the crack starts at the fusion line, Type III if the crack locates in the coarse grained HAZ (CGHAZ) and Type IV if the crack is in the fine grained HAZ (FGHAZ) or the intercritical zone (ICHAZ) [67].

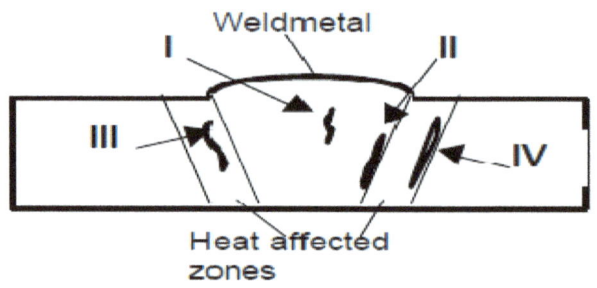

Fig. 2.54: Classification of crack locations [67].

Quantitative evaluation of creep damage is required in the heat-affected zones (HAZs) of welds in the main steam and reheat piping of fossil power boilers subject to internal pressure over a long duration [68]. Damage in the coarse-grain HAZ (Type III), and the fine-grain HAZ (Type IV), compose the main focus, with the latter, Type IV damage, being more important as the maximum damage occurs not on the surface but subsurface on piping. Figure 2.55, schematically illustrates the relationship between the number density of small defects and the creep-rupture time fraction, together with the progression of microscopic damage in the coarse-grain HAZ (CGHAZ hereafter) and the fine-grain HAZ (FGHAZ hereafter) at prescribed creep-rupture time fractions of 50%, 60% and 75%. In the Type IV damage mode, cavities or small defects with an average size of 10.5μm (the same size as the grain) initiate and coalesce together. In the Type III damage mode, on the other hand, cavities or small defects with an average size of 4μm initiate and coalesce on the boundary of grains with an average size of 50~200μm. In the final stage of creep rupture life, these defects grow into small cracks or crack-like defects, with a length of 0.1mm~1mm, throughout the wall thickness of the piping, and can lead to final failure of piping having a thickness of up to 100mm. Type III and Type IV damage, e.g. multi-site creep damage, is basically "initiation type". However, it also has the characteristics of "growth type" in the final stage of creep life as the growth of crack-like defects, in the high stress region of the weld thickness, governs the final failure life of the piping.

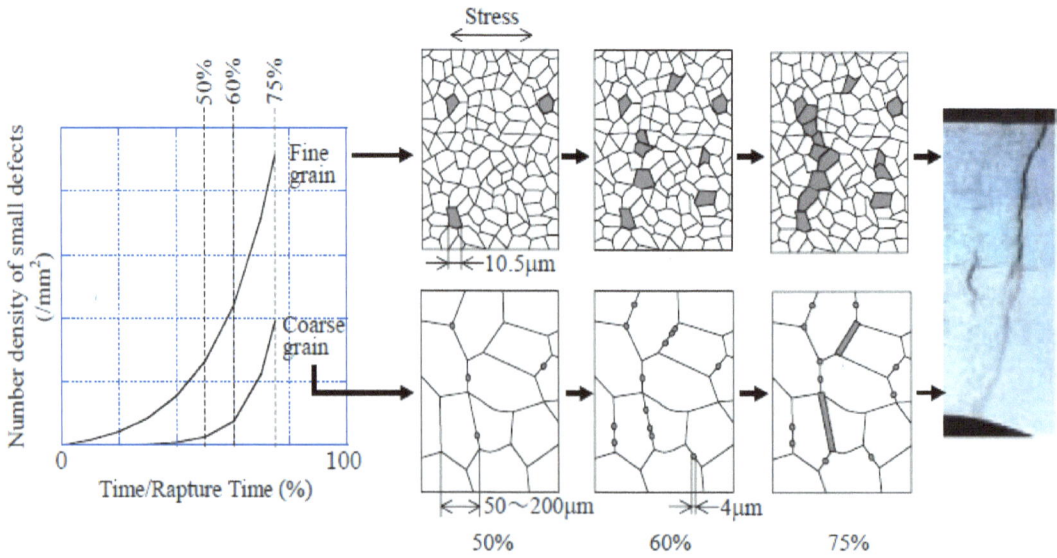

Fig.2.55: Schematic view of damage progress

Poor creep strength of fine-grained HAZ structure without lath-martensite is considered as the primary cause of Type IV fracture [69]. Further, grain boundary precipitates such as $M_{23}C_6$ and Laves phase, which are the nucleation site of creep voids, coarsen faster in fine-grained HAZ than in base metal during creep [69]. In order to improve the creep strength of fine-grained HAZ and resistance to Type IV cracking, strengthening of grain boundaries is considered to be efficient.

A proper addition of boron is generally considered to improve the creep strength of base metal by strengthening grain boundaries. It is reported that boron distributes within the grain boundary carbides, $M_{23}C_6$, and retards coarsening of them . 9Cr-3W-3Co-VNb steel with higher boron and low nitrogen content, which has better creep strength in long-term tests, has been developed at NIMS. In this steel, the boron addition also stabilizes the $M_{23}C_6$ precipitates and thus retards their coarsening during high temperature exposure. Therefore, it is expected that the creep strength of the fine-grained HAZ of welded joints can be improved by proper boron addition [69].

2. 9. 6. Factors affecting resistance to creep fracture:

Alloy additions in steel can significantly improve the resistance to creep fracture. Strong particles located at grain boundaries may hinder grain boundary sliding and therefore the development of intergranular cracks. Furthermore, as well as affecting creep rate and rupture life, alloying can also reduce creep ductility. For example, heat treatment procedures that result in precipitate free zones at grain boundaries can lead to low ductility by promoting the formation and propagation of intergranular cracks. In producing creep resisting alloys, with enhanced creep ductility, alloys have been produced with grains elongated in the tensile stress direction. Cracks form preferentially on boundaries oriented at right angles to the tensile axis. Crack link-up to cause failure occurs more easily with an equiaxed grain structure than with an elongated grain structure [70].

Generally in creep resistant alloy steels such as P91, solid solution strengthening of the matrix is an important factor. Such a matrix is more creep resistant compared to a pure metal as the elements in solid solution inhibit movement of dislocations through the crystal lattice. In addition to solid solution hardening, precipitation and work hardening can also effect a significant increase in the creep resistance. However, these hardening mechanisms are unstable relative to a rise in temperature. Therefore their use for the purpose of increasing the creep resistance of the metals is limited to the temperature ranges within which the strengthening mechanisms are stable [71].

The strengthening of a creep resistant alloy by a finely dispersed precipitate is subject to the same softening processes that occur in a normal precipitation hardening alloy. Coarsening and over-aging will occur with time at elevated temperature. Also heating of the alloy above the Ac3 temperature can cause re-solution of the precipitates in creep resistant steels. However, In welding, and the HAZ region is normally more prone to creep damage than the parent structure; and this discussed already through paragraph 2.9.5.2.

In fact, for a successful service application and acceptance in practice, the weldability and the long time behavior of the material is one of the most important aspects.

Chapter 3 : Experimental Work

3. 1. Experimental procedure:

The experimental work program presented through this research is summarized through Table 3.1; as following:

Table 3.1- Summary of the experimental work program presented through this research.

Objective
Investigation on the influence of welding parameters (Heat input) and heat treatments (Normalizing followed by tempering; compared with tempering condition alone) on weldments properties of Power Boiler Steel P91 (9Cr-1Mo-V-Nb).
First Step
Samples
Three P91 pipes of 323.9 mm outside diameter, 50.8mm thickness were welded with different heat inputs (1.15, 2.28 and 3.16 KJ/mm). All welds were intermediate post weld heat treated at 350°C to protect welds against cracking during handling and cutting **[73]**.
Second Step
Heat Treatment
One half of each welded joint was hence subjected to subcritical post weld heat treatment at 760 °C for 3.5 hrs with the same heating and cooling rates used during intermediate PWHT.
Third Step
Testing
<ul><li>Each samples subjected to the following tests and results were compared:</li><li>Tensile Test</li><li>Impact Test</li><li>Hardness Test</li><li>Stress rupture Test</li><li>Stress Corrosion Cracking Test</li><li>Optical Microscopic</li><li>SEM / X-ray Diffraction.</li></ul>

3. 1. 1. Materials:

The P91 steel used in this research was supplied by K & I Tubular Corporation as a pipe of 323.9 mm outside diameter, 50.8mm thickness and 6.5 m length. The P91 pipe steel was fabricated by hot forming followed by normalizing at 1050 °C for 10 min., and tempering at 785 °C for 45 min. Tables 3.2 and 3.3 show the chemical analysis and the mechanical properties of the as-received P91 pipe.

Table 3.2 — Chemical Composition for the as received P91 steel compared to material standard, ASME SA355 [74].

	Chemical Composition, wt%							
Element	C	Si	Mn	P	S	Ni	Cr	Mo
Standard (ASME SA335)	0.07-0.14	0.2-0.5	0.3-0.6	0.02 max	0.01 max	0.4 max	8.0-9.5	0.85-1.05
Actual	0.11	0.26	0.44	0.012	0.002	0.17	8.5	0.88
Element	Al	V	Nb	N				
Standard (ASME SA335)	0.04 max	0.18-0.25	0.06-0.1	0.03-0.07				
Actual	0.002	0.215	0.077	0.04				

Table 3.3: Mechanical properties and microstructure of the as received investigated steel P91.

Property					
Tensile Test			Average Hardness	Average V-Impact	Microstructure
Y.S, MPa	U.T.S, MPa	El, %	HB	J	Optical
501.3	669.5	49.5	207	224	Tempered Martensite

Three pipes were welded using different heat inputs (1.15, 2.28 and 3.16 KJ/mm). These heat input values were calculated based on SMAW arc efficiency of 0.8. In the actual girth welds of the P91 pipes, GTAW filler rod ER90S-B9 was used for root pass, while E9015-B9 SMAW electrode was used for first pass and filling passes. Table 3.4 summarizes the chemical composition of the used consumables. It should be noted that the root, first and cap passes were discarded during sample preparation for the different conducted tests.

Table 3.4: Chemical composition of GTAW rod ER90S-B9 and SMAW electrode E9015-B9 used during welding, wt%.

Electrode/Rod	C	Si	Mn	P	S	Ni	Cr
ER90S-B9	0.09	0.2	0.49	0.004	0.003	0.66	8.7
	Cr	**Mo**	**Cu**	**Al**	**V**	**Nb**	**N**
	8.7	0.9	0.03	0.006	0.19	0.08	0.05
E9015-B9	0.1	0.19-0.38	0.66-0.69	0.01	0.005-0.009	0.27-0.75	8.9-9.3
	Cr	**Mo**	**Cu**	**Al**	**V**	**Nb**	**N**
	0.9-0.98	< 0.1	0.001-0.002	0.21-0.26	0.04-0.06	0.03-0.04	----

Table 3.5 summarizes the welding procedures used during welding, and; the chemical composition of the used consumables.

Table 3.5. Summary of welding procedures used during pipes welding.

Process/ Layer no.	Current		Average Voltage	Average Starting Temp., °C	Average Heat Input kJ/mm	Filler Metal / Electrode	
	Type/ Polarity	AMP. Average				Type	Dia. (mm)
Sample No. 1 (Low Heat Input)							
GTAW (Root)	DC / EN	105.7	12.7	210	1.85	ER90S-B9	2.4mm
SMAW (1)	DC / EP	137	24	235	1.03	E9015-B9	2.5 mm
SMAW (2 to 13)	DC / EP	137	24	235	1.15	E9015-B9	3.25mm
Total No. of Passes = 100							

Process/ Layer no.	Current		Average Voltage	Average Starting Temp., °C	Average Heat Input kJ/mm	Filler Metal / Electrode	
	Type/ Polarity	AMP. Average				Type	Dia. (mm)
Sample No. 2 (Medium Heat Input)							
GTAW (Root)	DC / EN	106.8	11.08	220	1.7	ER90S-B9	2.4mm
SMAW (1)	DC / EP	81.5	21.5	230	1.08	E9015-B9	2.5 mm
SMAW (2 to 21)	DC / EP	135	25	225	2.28	E9015-B9	3.25mm
Total No. of Passes = 51							

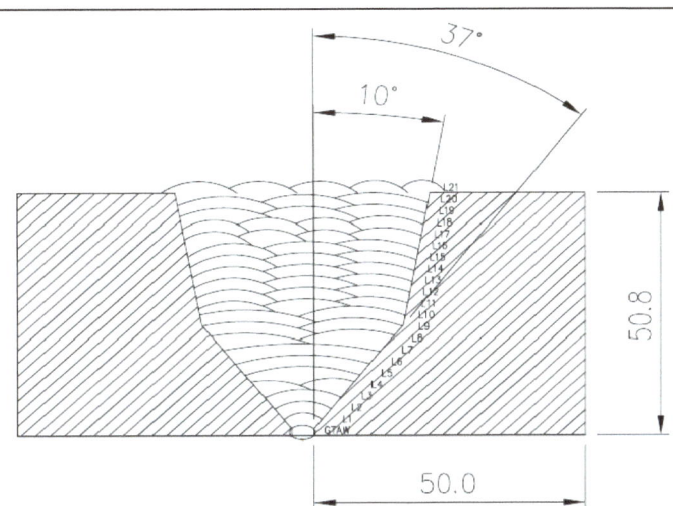

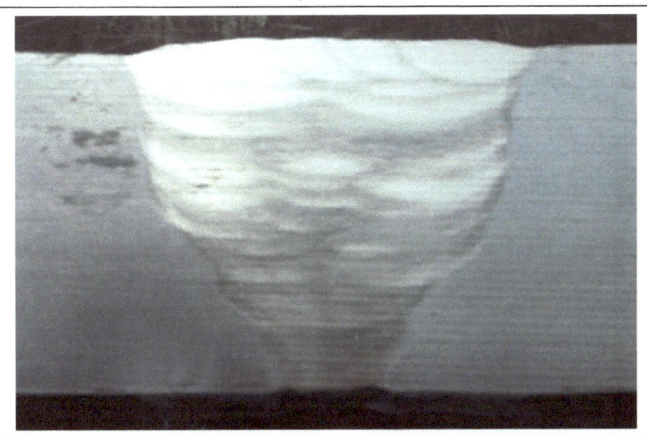

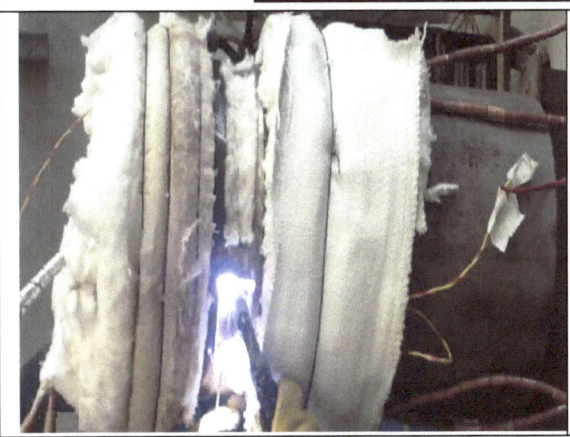

Process/ Layer no.	Current		Average Voltage	Average Starting Temp., °C	Average Heat Input kJ/mm	Filler Metal / Electrode	
	Type/ Polarity	AMP. Average				Type	Dia. (mm)
Sample No. 3 (High Heat Input)							
GTAW (Root)	DC / EN	92.5	1.18	210	1.32	ER90S-B9	2.4mm
SMAW (1)	DC / EP	82.3	21.6	210	0.92	E9015-B9	2.5 mm
SMAW (2 to 15)	DC / EP	160	25	210	3.16	E9015-B9	3.25mm
Total No. of Passes = 32							

73

3. 1. 2. Applied preheat and heat treatment:

3.1.2.1. Preheat and intermediate post weld heat treatment:

Preheating and intermediate post weld heat treatment were conducted using a calibrated portable heat treatment machine with ceramic blanket heaters (Fig. 3.1). Figure 3.2, illustrates the process steps starting with preheating (200°C – 250°C), control of inter pass temperature at 300°C and ending with the intermediate post weld heat treatment for 2 hrs at 350°C; this condition is considered as the as-welded condition.

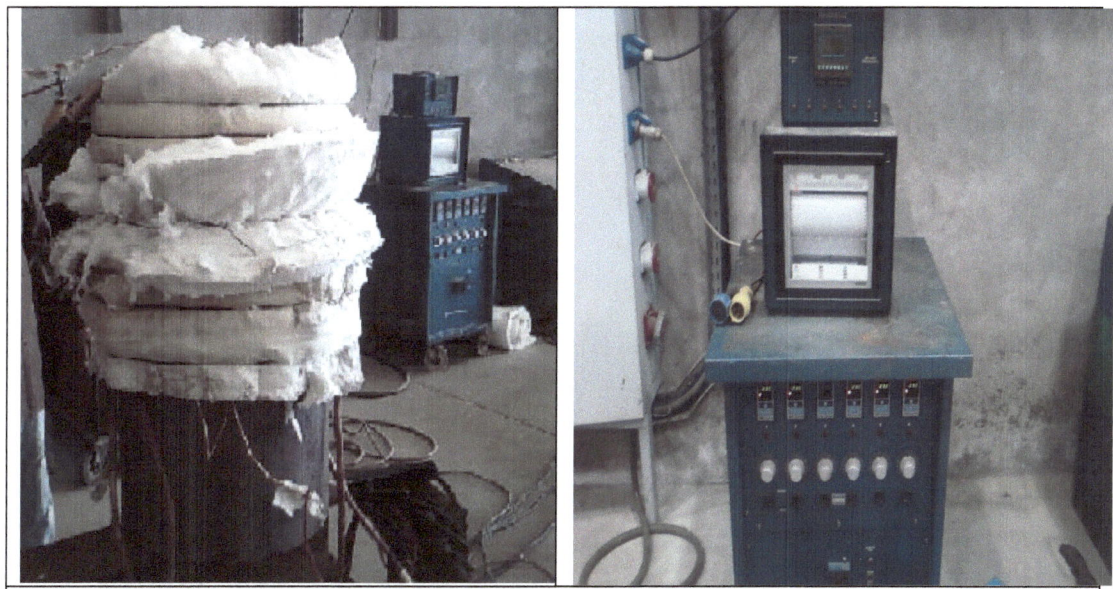

Fig. 3.1: Portable heat treatment machine with ceramic heaters used for preheating and intermediate post weld heat treatment.

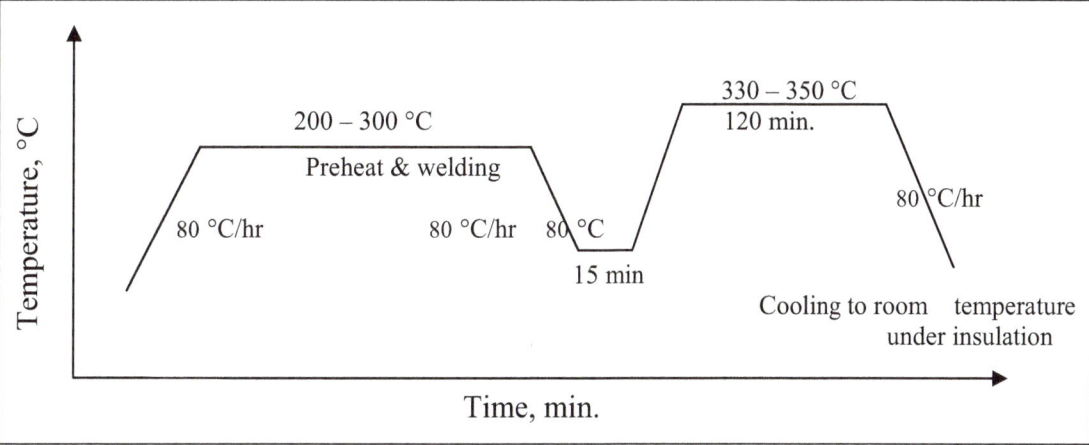

Fig. 3.2: Intermediate post weld heat treatment applied on each pipe after welding.

3.1.2.2. Heat treatment:

After intermediate post weld head treatment; each pipe was saw cut to several strips (see figure 3.3). However, as described previously through paragraph 3.1; two types of heat treatments have been applied for each heat input. One half of each welded joint was subjected to subcritical post weld heat treatment at 760 °C for 3.5 hrs with the same heating and cooling rates used during intermediate PWHT(see Fig 3.4). This heat treatment was suggested based on preliminary results presented in EPRI report 1004702 **[75]** which showed that the 760 °C subcritical PWHT temperature provides better tensile, Charpy, and rupture strength results compared to heat treatment at 649°C or 704°C The other halves of the welded coupons were subjected to a new heat treatment consisting of normalizing at 1050°C for 0.5 hr followed by tempering at 760°C for 3.5 hr as shown in Fig 3.4. The furnace used for tempering and normalizing/tempering heat treatments processes is shown in Fig 3.5. This furnace is equipped with control of heating and cooling rates.

Fig. 3.3: Water/Oil cooling saw cutting of welded pipes – First step of samples preparation.

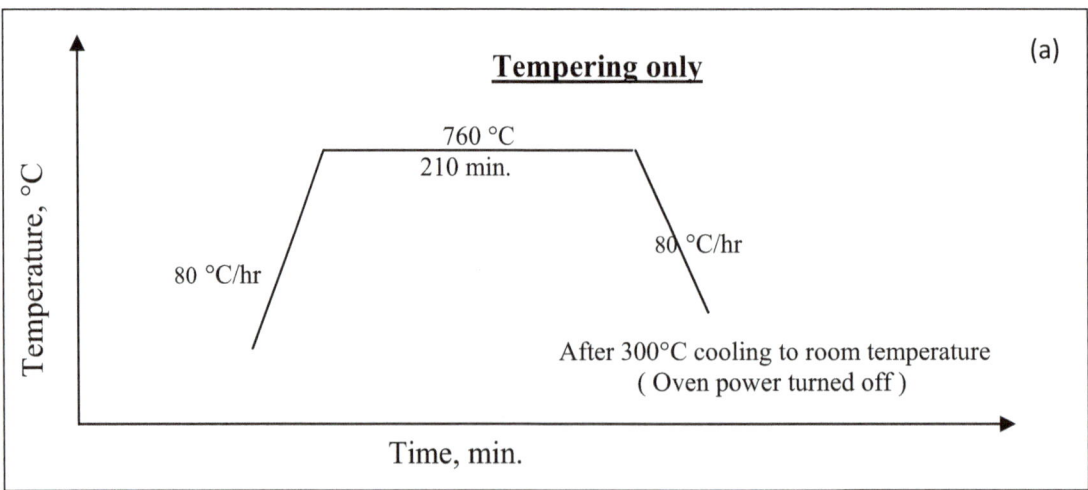

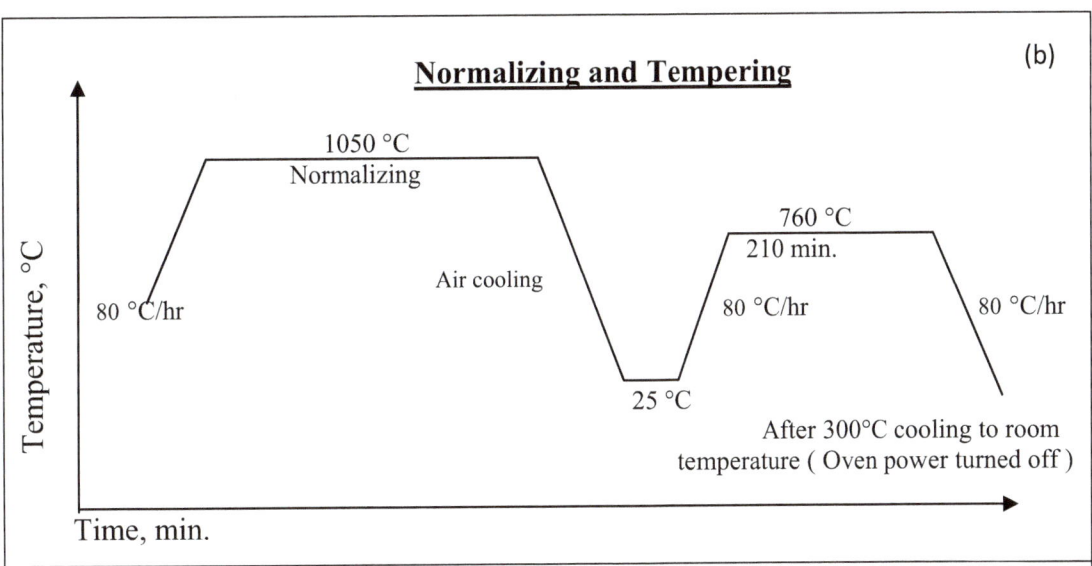

Fig. 3.4: Different post weld heat treatments applied on a part of each pipe after welding. (a) Tempering. (b) Normalizing and Tempering.

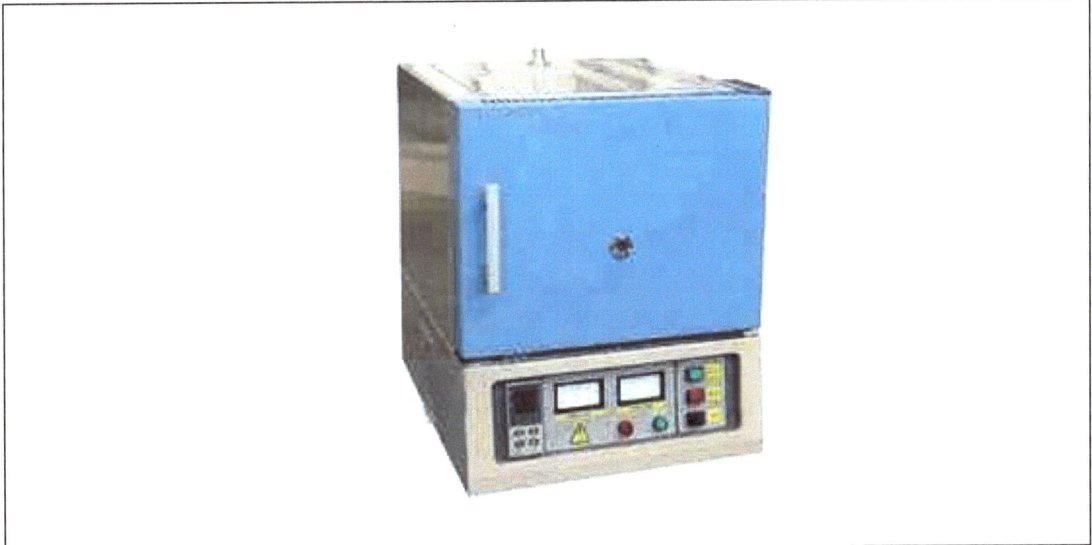

Fig. 3.5: Oven used for tempering and normalizing/tempering heat treatments processes with heating/cooling rates control.

3. 1. 3. Samples traceability and coding system:

For the sake of traceability of the various samples of different heat input/heat treatment conditions a coding system described in Table 3.6 was used throughout the present work.

Table 3.6: Coding system for different sample condition (Heat Input/ Heat Treatment).

Sample Code	Condition
L 350	Low heat input condition followed by intermediate PWHT at 350 °C for 2 Hrs. (As welded low heat input condition)
L 760	Low heat input condition followed by PWHT at 760 °C for 3.5 Hrs
L 1050	Low heat input condition followed by normalizing at 1050 °C for 0.5 Hr and tempering at 760 °C for 3.5 Hrs.
M 350	Medium heat input condition followed by intermediate PWHT at 350 °C for 2 Hrs. (As welded Medium heat input condition)
M 760	Medium heat input condition followed by PWHT at 760 °C for 3.5 Hrs.
M 1050	Medium heat input condition followed by normalizing at 1050 °C for 0.5 Hr and tempering at 760 °C for 3.5 Hrs.
H 350	High heat input condition followed by intermediate PWHT at 350 °C for 2 Hrs. (As welded high heat input condition)
H 760	High heat input condition followed by PWHT at 760 °C for 3.5 Hrs.
H 1050	High heat input condition followed by normalizing at 1050 °C for 0.5 Hr and tempering at 760 °C for 3.5 Hrs.
B 350	Base Metal treated by intermediate PWHT at 350 °C for 2 Hrs. (As-Received).
B 760	Base Metal treated by PWHT at 760 °C for 3.5 Hrs.
B 1050	Base Metal treated by normalizing at 1050 °C for 0.5 Hr and tempering at 760 °C for 3.5 Hrs.

3. 1. 4. Applied testing methods:

The influence of Heat input and different heat treatment (Normalizing followed by tempering compared with tempering condition alone) on microstructure, mechanical and corrosion properties of Power Boiler Steel P91 (9Cr-1Mo-V-Nb) have been studied by means of the following tests:

3. 1. 4. 1. Tensile test:

Tensile test was carried out using calibrated universal tensile machine having a digital indicator with three different ranges (0 to 6 tonf, 0 to 30 tonf and 0 to 60 tonf). The test was carried out using 0 to 6 tonf range. Cylindrical test specimens shown in figure 3.6 and 3.7 were machined from investigated material. Two samples from each condition were tested and the average value was reported.

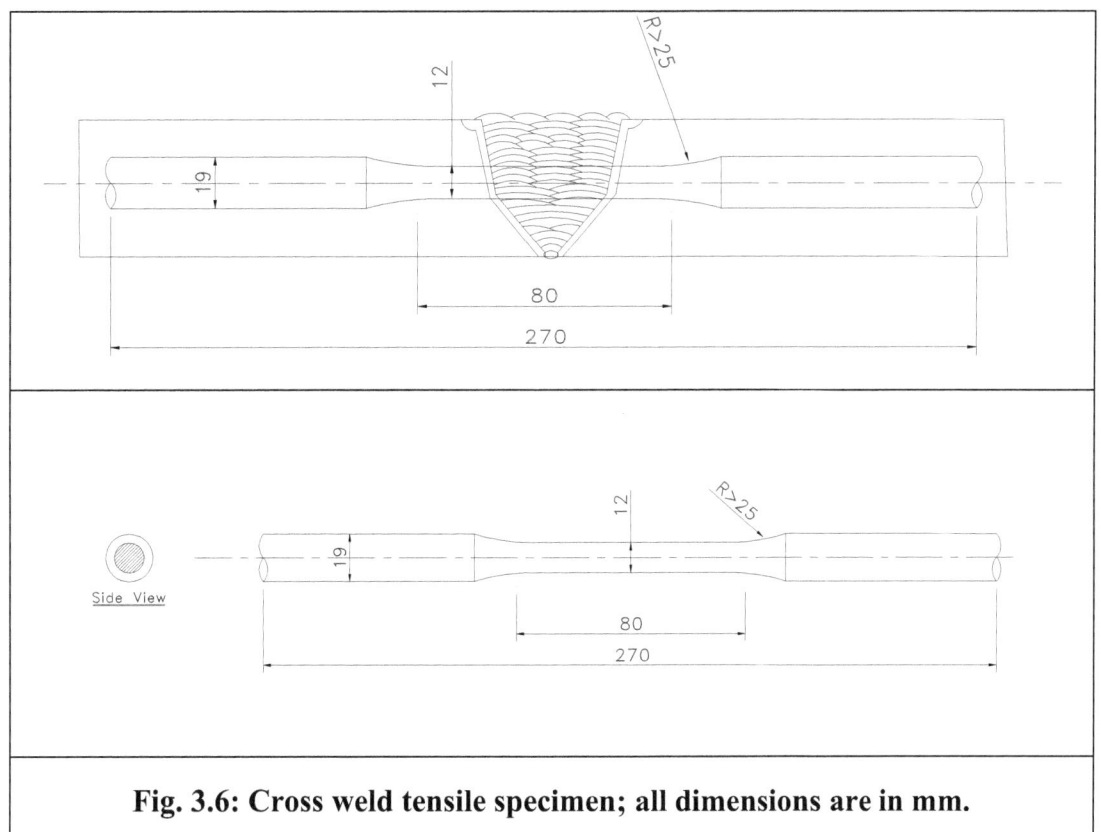

Fig. 3.6: Cross weld tensile specimen; all dimensions are in mm.

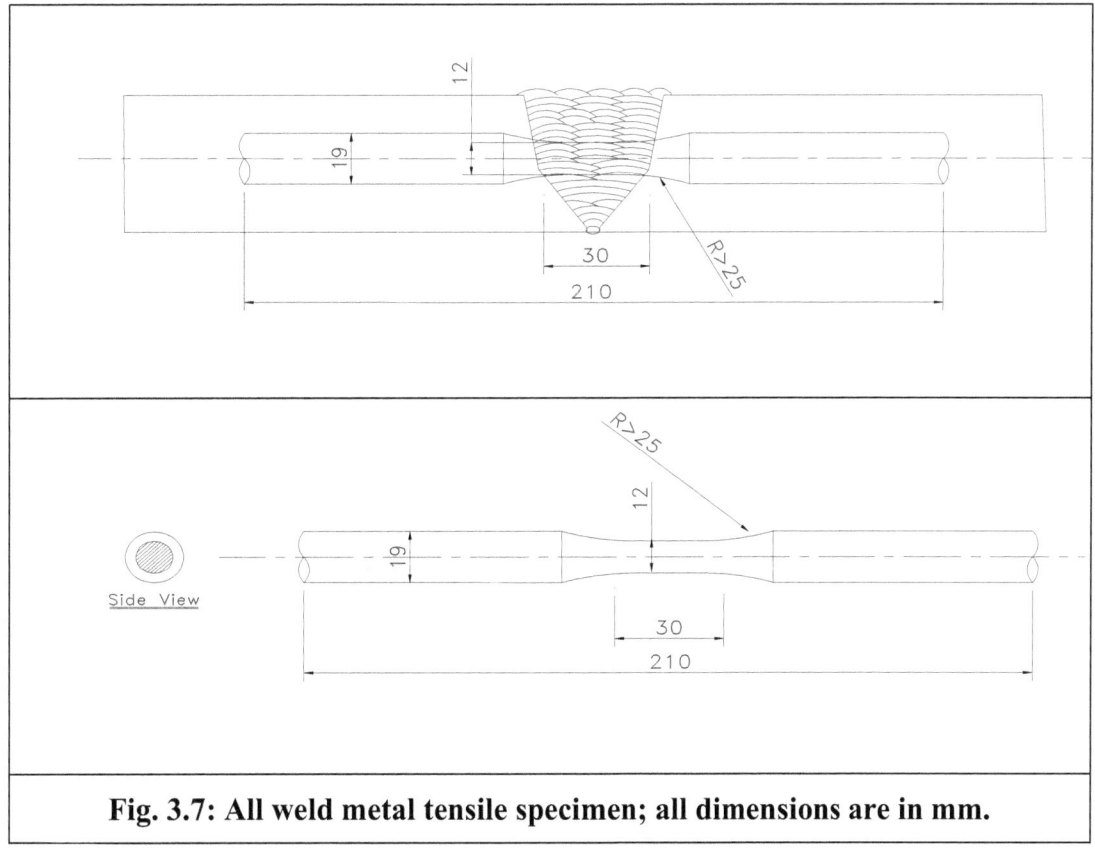

Fig. 3.7: All weld metal tensile specimen; all dimensions are in mm.

3. 1. 4. 2. Impact test :

The test were performed at ambient temperature on a standard sample of 10 mm square section x 55 mm; with a 2 mm deep V-notch positioned at the middle of the specimen and located to cover each area of the welded joint, see figure 3.8. All specimens were fractured using a 500 Joule pendulum hammer. Three sets covering the weld, HAZ and base metal of each condition were tested. The three impact specimens of each set were tested and the average value was reported.

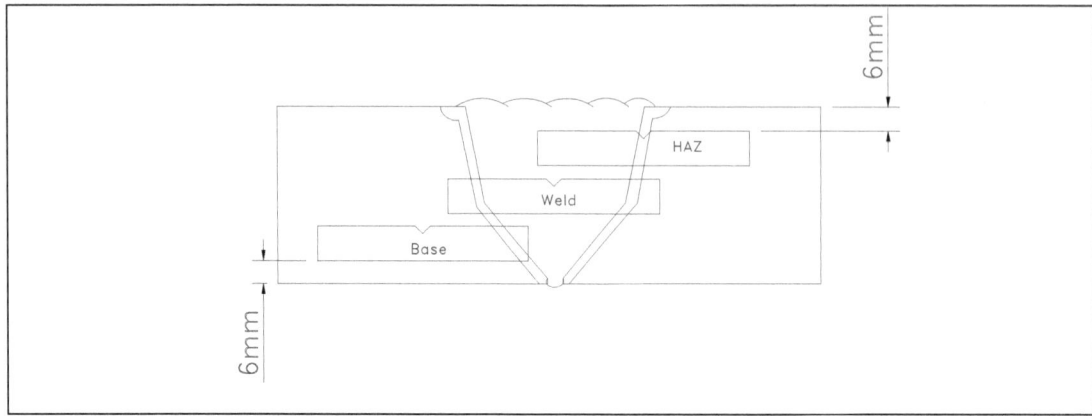

Fig. 3.8: Direction and location of V-notch of the impact specimen.

3. 1. 4. 3. Hardness test:

Vickers hardness measurements were made by a bench hardness testing machine using a load of 10 kg (HV 10). Tests were performed on the same flat cross-sectional specimens used for optical microstructure examination; and the measurements started from weld and go through HAZ and base metal to give the actual profile of hardness distribution along different areas of welded joint.

3. 1. 4. 4. Stress Rupture Test:

Constant load uniaxial stress rupture testing in tension were conducted on cross-weld cylindrical specimens shown in figure 3.9. Test was performed using a calibrated lever arm creep test machine fabricated by Project Services Company with 2000 Ibf capacity and 1: 6.2 loading ratio (See figure 3.10). Four samples of each heat treated conditions were tested at 630°C, 650°C, 665°C and 677°C under stress of 135 MPa; and time to rupture, elongation and reduction of area were recorded. Specimen temperature was controlled by a thermocouple contacted to the specimen center.

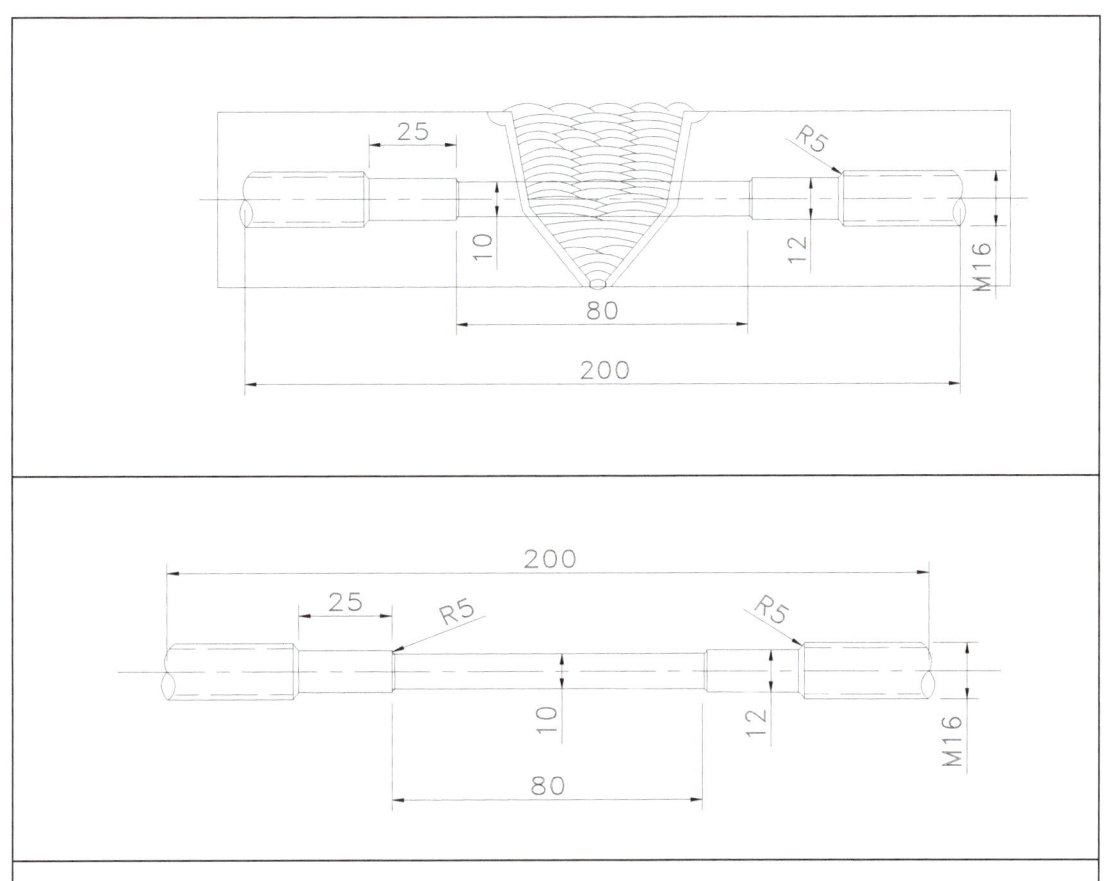

Fig. 3.9: Cross weld stress rupture specimen; all dimensions are in mm.

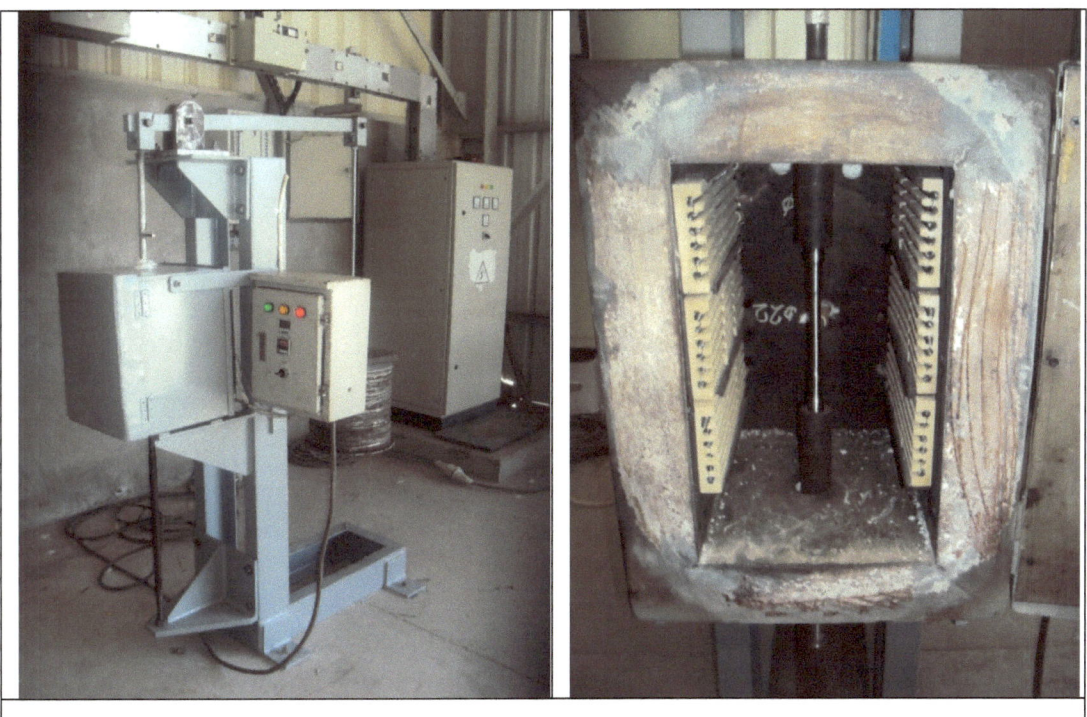

Fig. 3.10: Creep test machine used in the stress ruptures testing.

3. 1. 4. 5. Stress corrosion cracking test:

Fig. 3.11: Stress-corrosion test using U-bend specimens.

Stress corrosion cracking (SCC) in boiling magnesium chloride (155.0°C (311.0°F)) were used to compare and evaluate the weldments resistance to stress corrosion cracking. For most applications, this environment provides an accelerated method of ranking the relative degree of stress-corrosion cracking susceptibility for stainless steels and different alloys in aqueous chloride environment. Materials that normally provide acceptable resistance in hot chloride service may crack in this test. The preparation of test solution and testing procedure were conducted in accordance with ASTM G36 **[76]**. U-bend specimens in according to ASTM G30 were used **[77]**.

The U-bend specimens were prepared by bending a rectangular (5.5mm thickness by 20 mm in width) strip by an angle of 180° around a predetermined radius (17mm) and maintaining this constant strain condition during the stress corrosion test, see figure 3.11. SCC specimens were tightened using a 304 stainless steel bolt to maintain the 180° bent. The bent area in all specimens was selected to cover HAZ, weld and base metal; and the edge of each specimen was hard marked with its condition identification to assure of proper traceability. Periodic removal of the specimen, each 12 hrs, from the solution was necessary to determine the time when first crack appears. The outer surface of the U-bend was visually examined using a magnifying lens in order to detect crack initiation.

3. 1. 4. 6. Microstructure investigation:

Microstructure examination, for each condition, were conducted on a cross weld samples (25mm x 25mm) selected from the center of cross section of the welded joint. Each sample was then subjected to detailed microstructure examination using optical microscope, Micro chemical analysis using energy dispersion X-ray (EDS) and Scanning Electron microscope (SEM). On the other hand, other samples containing the weld metal, HAZ and base metal were selected from the impact specimens of each condition; and subjected to X-ray diffraction analysis to indicate different phases within the microstructure and to detect the presence of austenite phase if any.

3. 1. 4. 6. 1. Optical Microscope:

Specimens for optical microscopy were cut from actual welds and heat treated samples using a water/oil cooling saw cutting machine. The specimens were prepared for metallographic observation according to a standard metallographic sample preparation procedure. The samples were ground flat using 220 to 1000 SiC emery papers with water as lubricant and coolant. After grinding the specimens were polished with a Md-Mol cloth and 3μm diamond spray.

To reveal the microstructural constituents of the specimens with sufficient contrast, various etching reagents were employed, but the best results were obtained using Villella's Reagent (Table 3.7). Villella's etching solution revealed the different types of martensitic matrix (Laths or Plates), while the α-ferrite and δ-ferrite were revealed as white phases. On the other hand, Villella's etching solution, also, revealed the prior γ-grain boundaries as a white area.

Table 3.7: Chemical composition of Villella's Reagent.

Chemical composition	Type of etching
100 ml Ethanol 5ml HCl 1 g Picric acid (Picric acid is mixed with Ethanol then HCl is added).	Dip etching for 120s to 180s

After etching the specimens were rinsed with water, washed by methyl alcohol and finally dried in a stream of warm air; then viewed under a metallurgical optical microscope (model XJP-6A, Qualitest) using 100X, 200X and 400X magnifications.

3. 1. 4. 6. 2. Scanning electron microscopic (SEM) and Energy Dispersive X-ray (EDS):

Scanning electron microscopic (SEM) investigation and micro-chemical analysis using energy dispersive X-ray (EDS) were used to study the microstructure in more details. The investigation was conducted in the Egyptian Central metallurgical research and development institute (CMRDI); using JEOL JSM-5410 SEM with energy dispersive X-ray spectrometer, see figure 3.12.

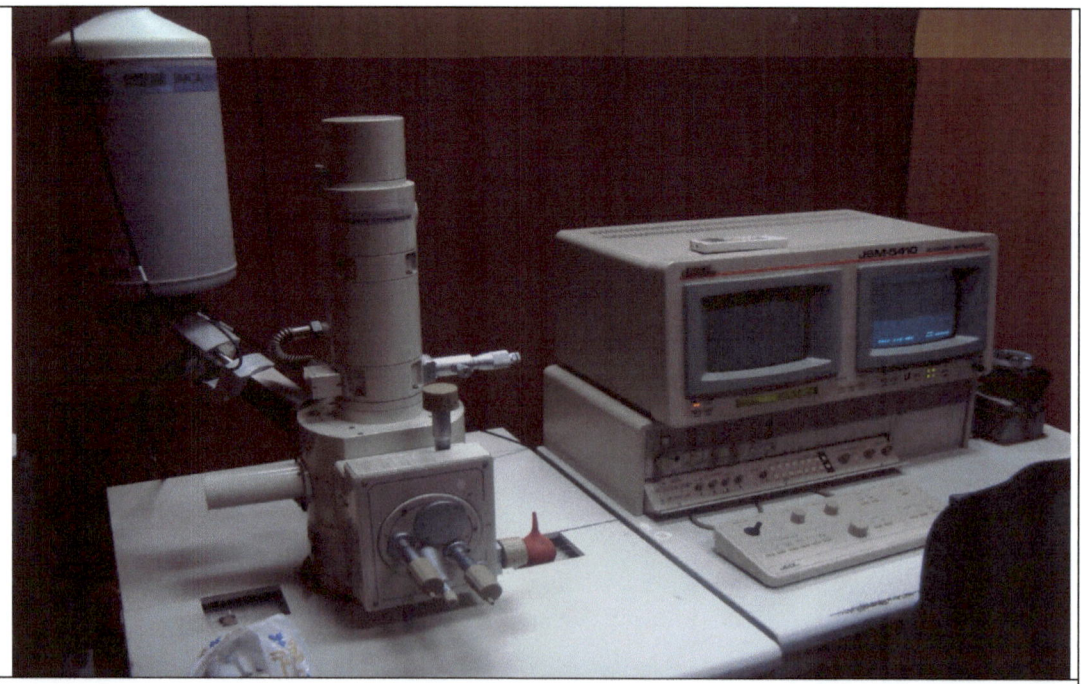

Fig. 3.12: JEOL JSM-5410 - Scanning electron microscopic (SEM) with energy dispersive X-ray spectrometer; used for both SEM and EDS analysis.

Both SEM and EDS investigations were conducted on the same samples that were used for optical microscopic examination with the same preparation and etching conditions.

3. 1. 4. 6. 2. 1. Roles of carbides identifications:

Most of carbides can, of course, be identified uniquely using electron diffraction, but this is very difficult in practice when examining large numbers of small particles. The carbides are often too thick or too thin to provide reasonable diffraction data, and the diffraction can sometimes be ambiguous [78].

A frequently used alternative method in this type of research is based on EDS microanalysis. This technique for carbide identification is carried out in the SEM and permit chemical analysis to be taken from individual carbides. Pilling and Ridley [79], found that the carbides showed one of five characteristics X-ray spectra as shown in figure 3.13.

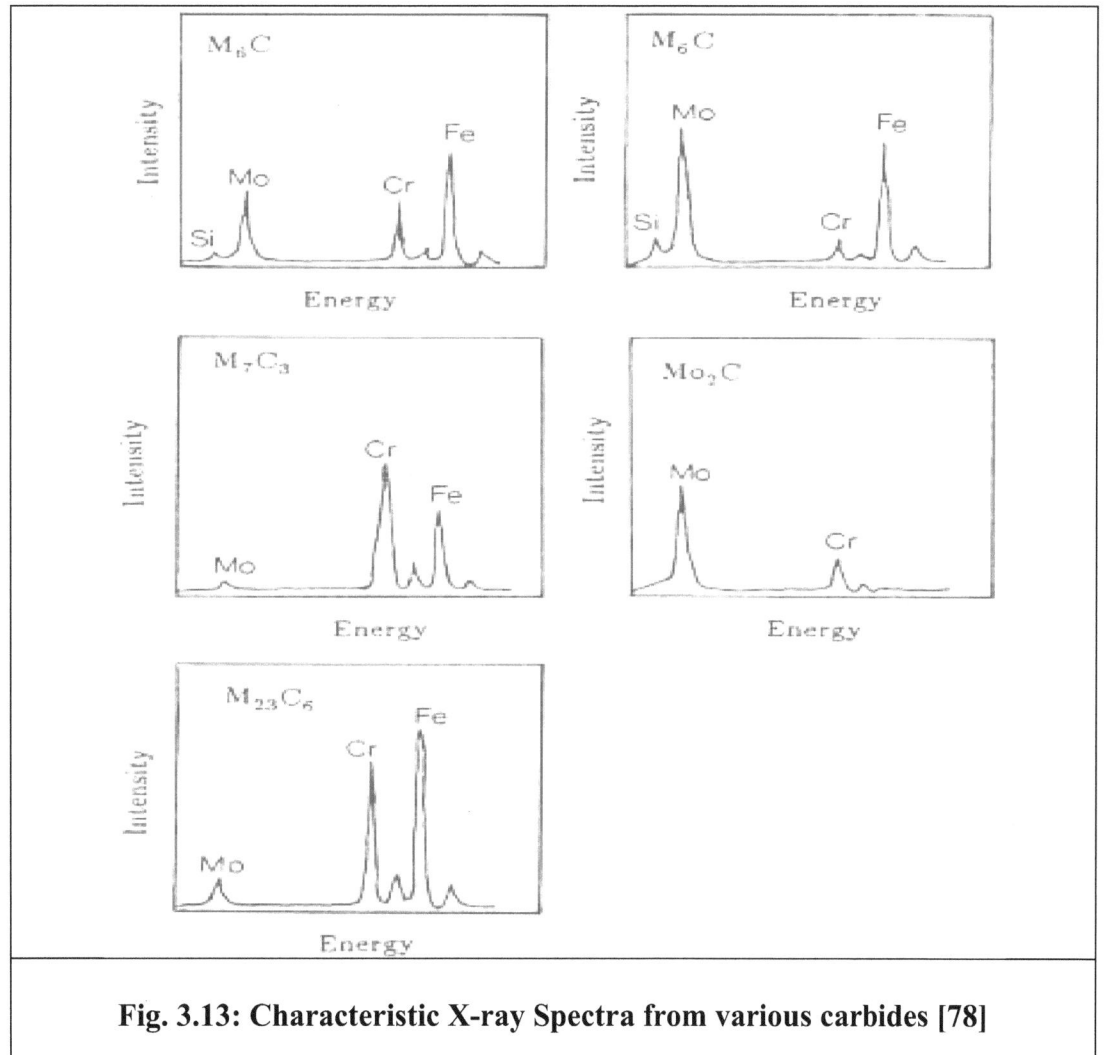

Fig. 3.13: Characteristic X-ray Spectra from various carbides [78]

Figure 3.14 show the characteristic X-ray spectra for Laves phase and VC as an example for MX precipitates in steel P91 **[82]**.

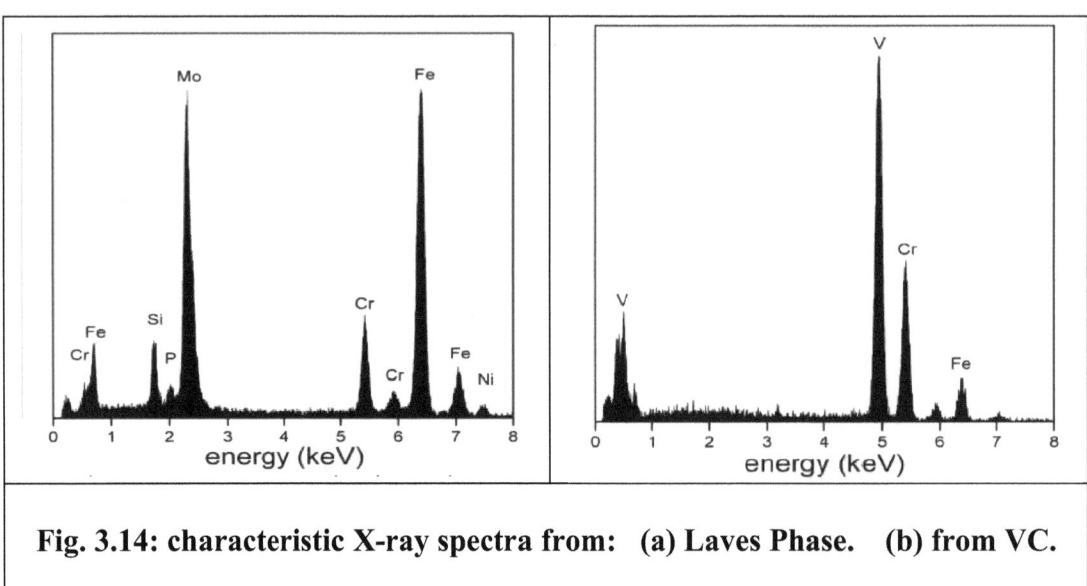

Fig. 3.14: characteristic X-ray spectra from: (a) Laves Phase. (b) from VC.

One of the important roles used for carbides identification by EDS technique is its chemical composition. Figure 3.15; illustrates the types of carbides identified and the distribution of major alloying elements in them **[83]**.

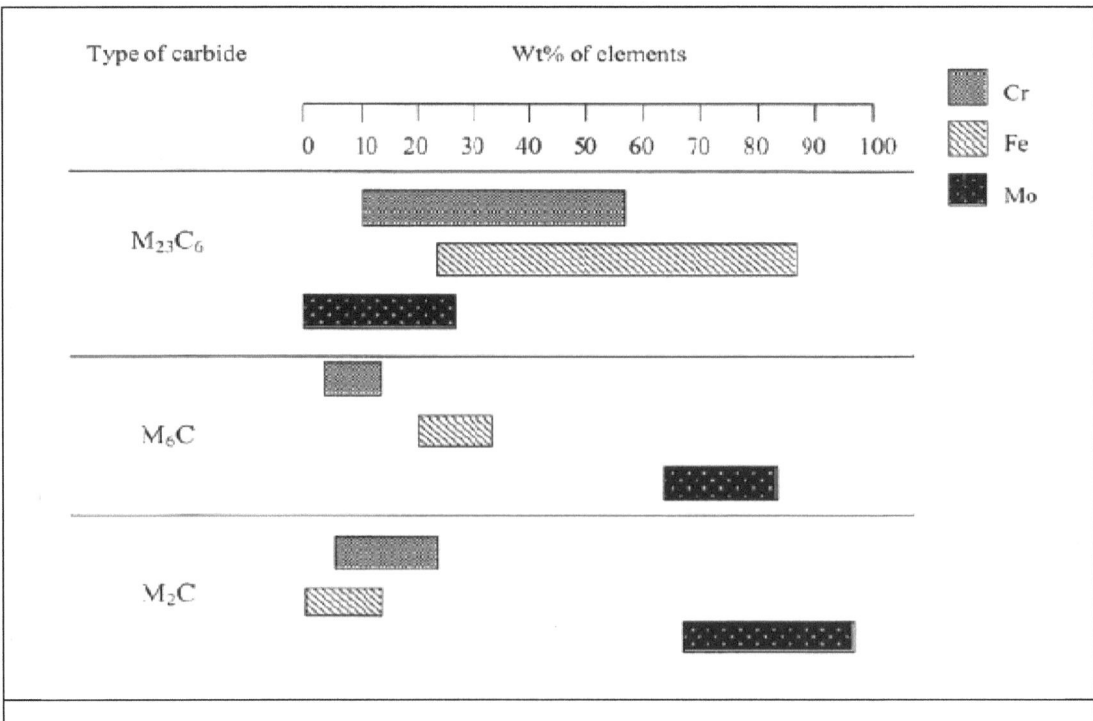

Fig. 3.15: illustrates the types of carbides identified and the distribution of major alloying elements in them [83].

Morphological differences can also act as a guide in the identification of carbides. Pilling and Ridley [79], suggest that there are four distinct carbide morphologies in specimens tempered at 700°C:

(i) Globular precipitates at the prior austenite and lath grain boundaries
(ii) Rod like precipitates in the matrix
(iii) Clusters of needle shaped precipitates in the matrix
(iv) Parallelogram shaped particles in the matrix.

The distinctive needle shaped morphology is adopted by Mo_2C. Pilling and Ridley [79], suggested that grain boundary carbides are M_6C or $M_{23}C_6$ and that the rod shaped precipitates are M_7C_3.

However, Balluffi et al [84], have shown that $M_{23}C_6$ has a spherical form and Beech and Warring [85], have shown that M_7C_3 and $M_{23}C_6$ have similar morphologies. Thus, it can be concluded that the shape of the particle alone is not a guide to identification.

All these methods, namely, electron diffraction, micro- analysis, and morphological observations were used to Identify the carbides.

3. 1. 4. 6. 3. X-ray diffraction:

X-ray diffraction analysis was performed using a diffractometer with a monochromatic source of radiation (CuKα), (PaNalytical, X'pertpro), and was carried out on each condition using the impact specimens. A sample from weld, HAZ and base metal were cut from each welding/heat treatment condition. The specimens were located in the instrument stage and scanned over the Bragg's angle (2Θ) range 40-120°. The diffraction angles were signified by peaks on the chart. The peaks were analyzed and the d-spacing was calculated from Bragg's law. Analysis of each peak was conducted using X' pertpro software and the best fit diffraction pattern and was compared with X-ray diffraction data cards of the international Center of diffraction data (ICDD); which are uploaded to the used software. The used software also measures the volume fraction of detected phases.

On the other hand, Martensite peaks were, additionally, identified by comparison with peaks obtained from the X-ray diffraction carried out on a water quenched sample of P91 steel. This sample was quenched in water from the same temperature used for normalization process (1050 °C). Samples were subjected to microstructure examination to ensure 100% martensitic structure, but actually a traces percentage of ferrite grains have been observed. Figure 3.16 and table 3.8 shows the XRD pattern and d-spacing for the martensitic sample.

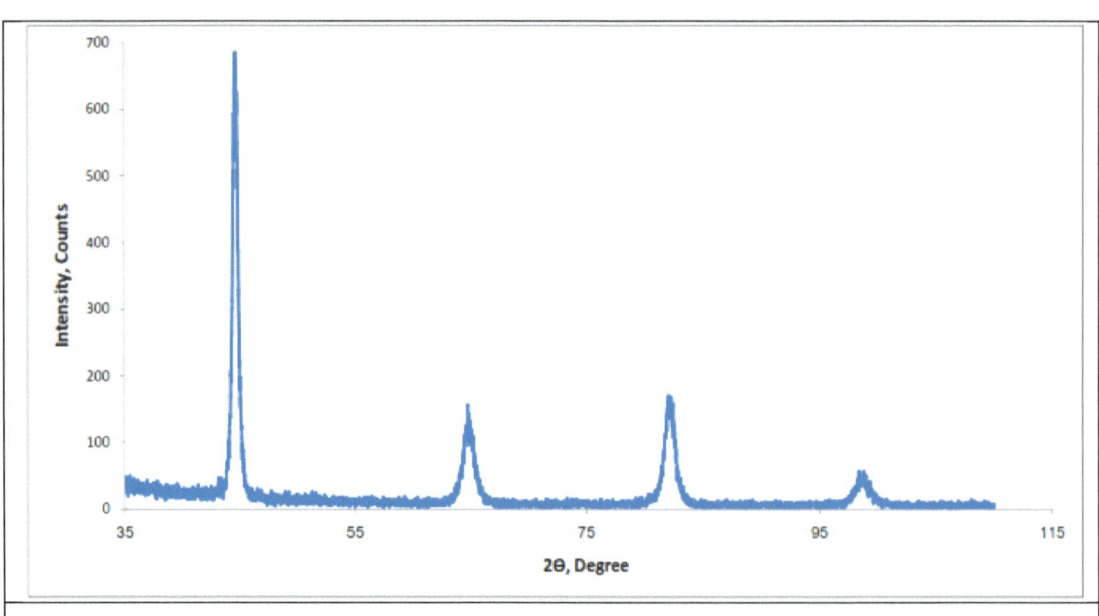

Fig. 3.16: X-Ray diffraction pattern of P91 base metal normalized at 1050 °C for 0.5 hr and quenched in water.

Table 3.8: XRD peaks analysis from the XRD pattern of the as quenched specimen (Standard Martensitic sample).

2Θ°	d-Spacing, Å	FWHM	Area[cts *°2Th.]	Back ground. [cts]	Height [cts]	Matched with Phase/(*hkl*)
44.583	2.03242	0.4330	286.69	10.00	671.26	Martensite, α & Bainite/(110)
64.636	1.44202	0.7872	83.59	3.00	107.65	Martensite, α & Bainite (200)
82.078	1.17418	0.9446	142.21	2.00	152.61	Martensite, α & Bainite/(211)
98.594	1.01609	1.1520	65.74	3.00	42.80	Martensite, α & Bainite/ (220)

3. 1. 4. 6. 4. Ferritescope examination:

This method was used to follow up the change in the magnetic permeability of the obtained microstructure. This gives estimation about the degree of carbides precipitation due to the decomposition of martensite during tempering treatments and its effect on the reduction in the carbon content of martensite that affects martensite hardness. It works also as a good indicator for comparing the expected volume fraction of martensite and ferrite in different specimens of different welding/heat treatment conditions. The ferrite content of all test specimens was measured more than three times and the average value of these readings was reported. The measurements covered different areas of the weldments (HAZ, Weld and Base metal). Ferrite measurements were carried out using Fischer MP 30 Ferritescope.

Chapter 4 : Results and Discussion

4. 1. Mechanical Properties:

4. 1.1. Hardness testing results

Fig. 4.1 shows the hardness profiles along weldments for various heat input and heat treatment conditions. The highest hardness was obtained at the weld metal of the as-welded conditions and its value decreases with the increase in the heat input (see Fig 4.1(a) for weld metal). The maximum hardness obtained within the weld metals came as a result of the formation of un-tempered martensite microstructure. The hardness of the formed untempered martensite is expected to decrease with the increase of heat input due to the reduction in cooling rate from welding temperature, as per the existed welding condition, which enhances the diffusion and migration of carbon atoms out of martensite grains, and enhances the carbides precipitation within the matrix. This supported be the resulted ferrite content measured by the ferrite scope in the weld zone which indicated the increase of martensite magnetic permeability with heat input, 75%, 77.5% and 82% respectively. However, magnetic permeability of martensite increases as the carbon concentration within the martensite grains decreases.

In general, the hardness in HAZ starts to decrease from the fusion line toward the base metal, but two important phenomena were observed: The hardening effect at the Coarse grain heat affected zone (CGHAZ) just beside the fusion line as a result of complete dissolution of precipitates which enhanced the grain coarsening and the formation of high carbon and hard martensite, and the softening effect that started at the fine grain HAZ (FGHAZ) where partial dissolution of precipitates leads to the formation of fine grain structure and the formation of lower carbon and softer un-tempered martensite than that obtained at the CGHAZ. The softening effect continued through the Inter critical HAZ (ICHAZ) where the dissolution of precipitates and the amount of formed un-tempered martensite are at minimum. However, the major softening effect and the minimum hardness value were reached at the over tempered zone in which no dissolution of precipitates or formation of un-tempered martensite takes place, only over tempering of the base metal takes place. The hardness then increased again gradually till reaching a constant value at the un-affected base metal.

After the subcritical PWHT (Fig.4.1 (b)) the hardness of the weld metal decreased and became much lower than that of the as-welded conditions with the same trend; namely the greater the heat input the lower the hardness. On the other hand peaks of hardness appeared at the CGHAZ of both medium and high heat after tempering,

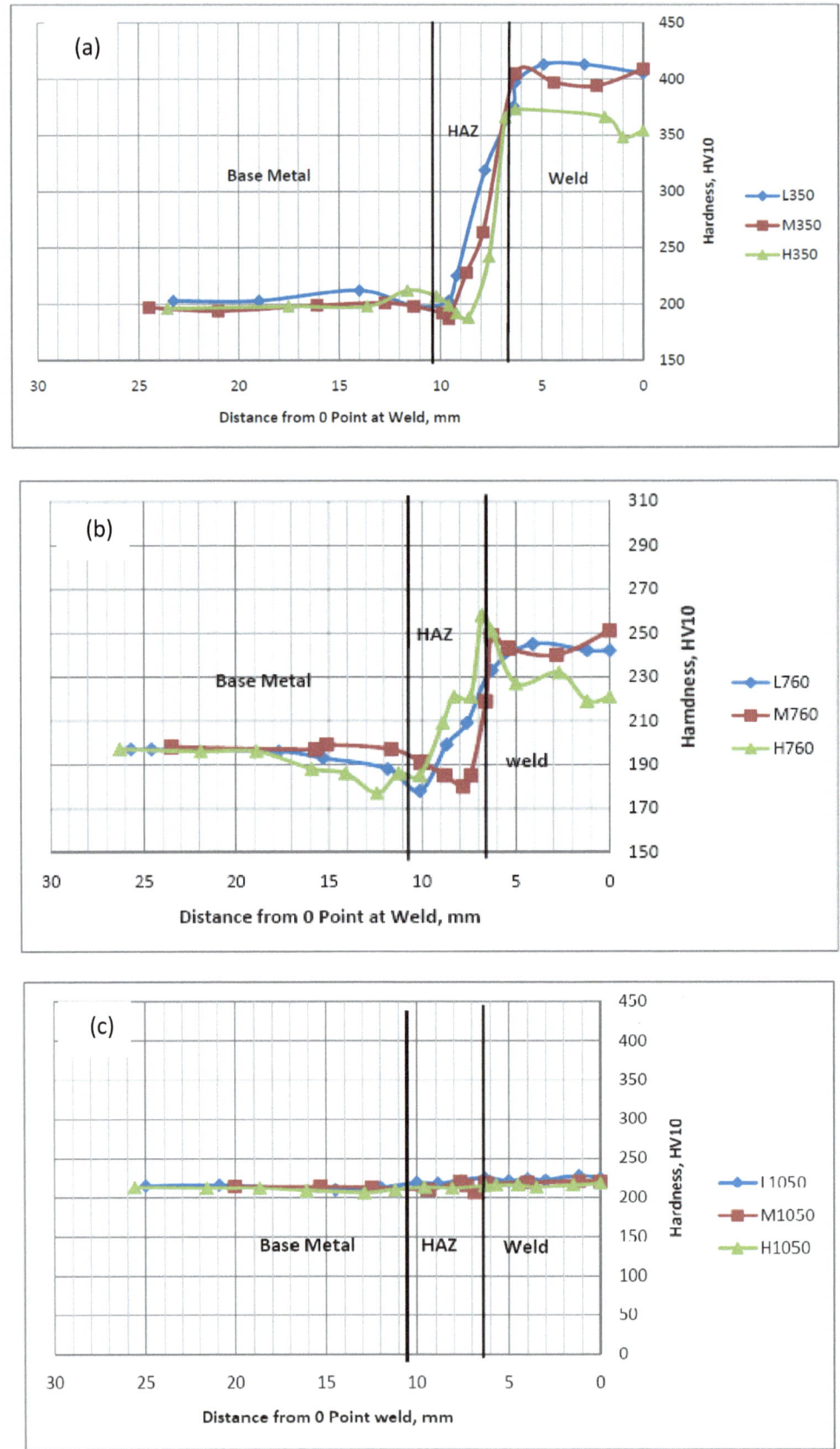

Fig. 4.1. Hardness profiles along weldements for various heat input conditions: (a) As-welded condition. (b) Subcritical PWHT. (c) N&T heat treatment.

The final concentration and distribution of carbon along the HAZ is shown to be directly depending on the temperature gradient and compositional differences in alloying elements along the heat affected zone which occurs as a result of the peak temperature produced at different location. For the HAZ, the highest peak temperature due to the thermal cycle during welding is located at the CGHAZ then the temperature decreases gradually toward the unaffected base metal.

Considering the high values of carbon mobility at the usual welding temperatures carbon redistribution can be expected during high-temperature exposure of lower alloy area of FGHAZ and higher alloyed CGHAZ. Carbon will diffuse in the direction of its thermodynamically activity gradient - disregarding the direction of the concentration gradient - and carbon enriched (CEZ) and carbon depleted zones (CDZ) will be formed in the weld area. Thus the FGHAZ becomes weaker and the CGHAZ becomes harder [85].

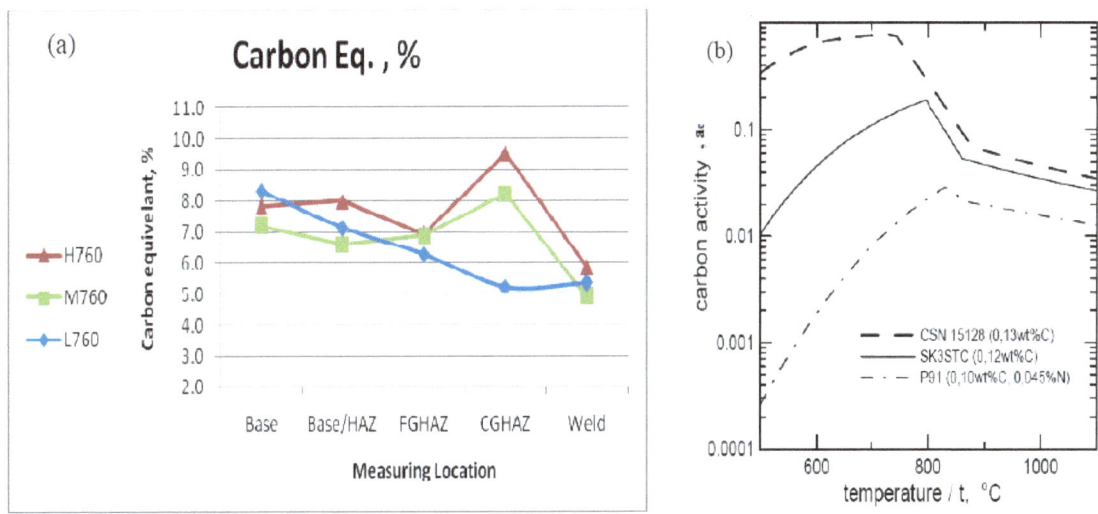

Fig. 4.2. (a) Carbone equivalent calculations along weldements of different heat inputs in the subcritical PWHT conditions; (b) Carbon activity, a_c, of the different steels included P91 at different temperatures[86].

The higher the peak temperature the higher the carbon activity, a_c, (The ability of carbon atoms to move by an interstitial or interstitialcy mechanism), (Fig. 4.2 (b)), which enhanced the diffusivity of carbon atoms and enhanced its migration from the ICHAZ and FGHAZ toward the CGHAZ [86]; and this leads to the higher concentration of carbon and alloying elements at CGHAZ and the lower concentration at FGHAZ (Fig.4.2 (a)) especially in the case of higher heat input. Dejun Li, based on SEM observation conducted on simulated HAZ of high Cr heat resistance steel after PWHT, reported that the precipitates in FGHAZ specimens were fewer and larger than those in CGHAZ specimens and base metal specimens [87].

Moreover, the softening effect in the FGHAZ became clearer and more significant after the subcritical PWHT. As a result of the re-tempering effect, the hardness of base metal affected by the subcritical PWHT is slightly lower than that obtained for the base metal in the as-received condition (the difference is about 10 HV10).

The hardness was nearly the same for different heat input conditions when the weldements were subjected to normalizing and tempering (N&T) treatment (Fig.4.1(c)). This is due to the high degree of microstructure homogeneity (almost similar microstructure of tempered martensite) in all investigated heat input conditions after normalizing and tempering treatment as a result of the phase transformation to austenite on heating which equalizes the microstructure in the weld, HAZ, and base metal in the form of fine grains of austenite.

However, after normalizing and tempering some differences remained in the in microstructure, Fig. 4.49 and 4.50. But the distribution of precipitates along the weldement became more homogeneous. Fig. 4.3 summarizes the results of EDS analysis conducted along the weldement of medium heat input after normalizing and tempering which gives an idea about the distribution of precipitates especially through the HAZ.

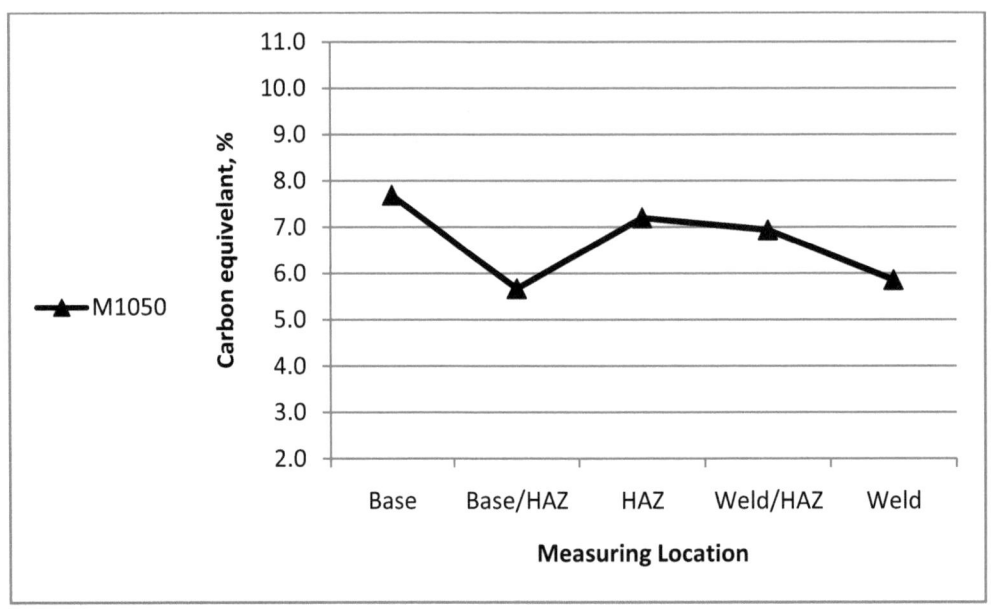

Fig.4.3. Carbone equivalent calculations along weldement of medium heat input after normalizing and tempering.

4. 1. 2. Tensile test results:

Fig.4.4 (a-b) shows the UTS values of specimens studied in this work. According to Fig.4.4(a) for the as welded condition, the tensile strength decreased with the increase in heat input during welding, as the lowest heat input condition resulted in the highest weld metal tensile strength. This is a result of the highest volume fraction of higher carbon and harder un-tempered martensite being associated with the lowest heat input. After subcritical PWHT, the tensile strength values were reduced to about 65% of that of as welded conditions as a result of the martensite decomposition during this tempering treatment.

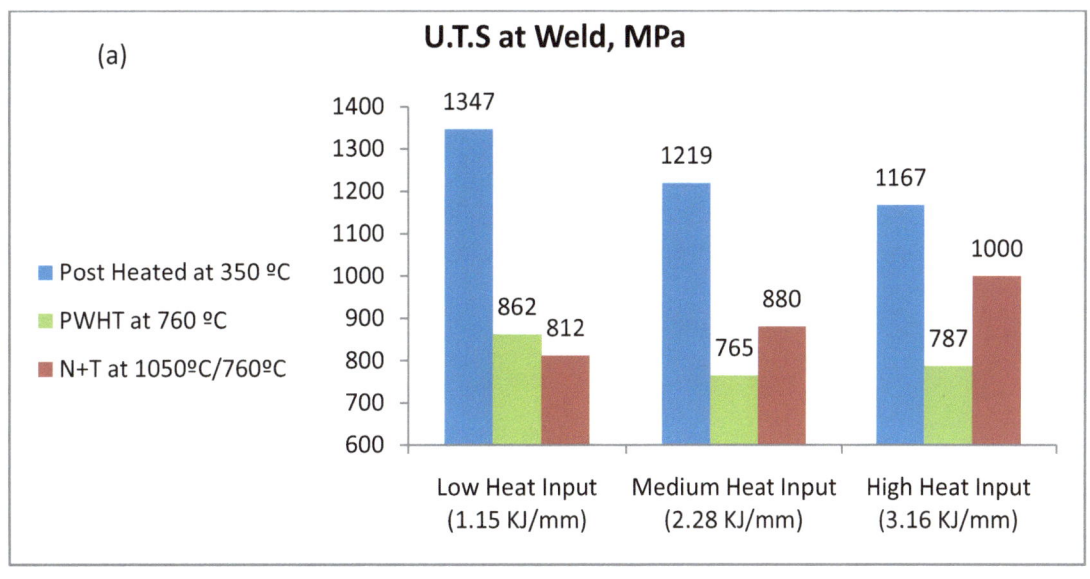

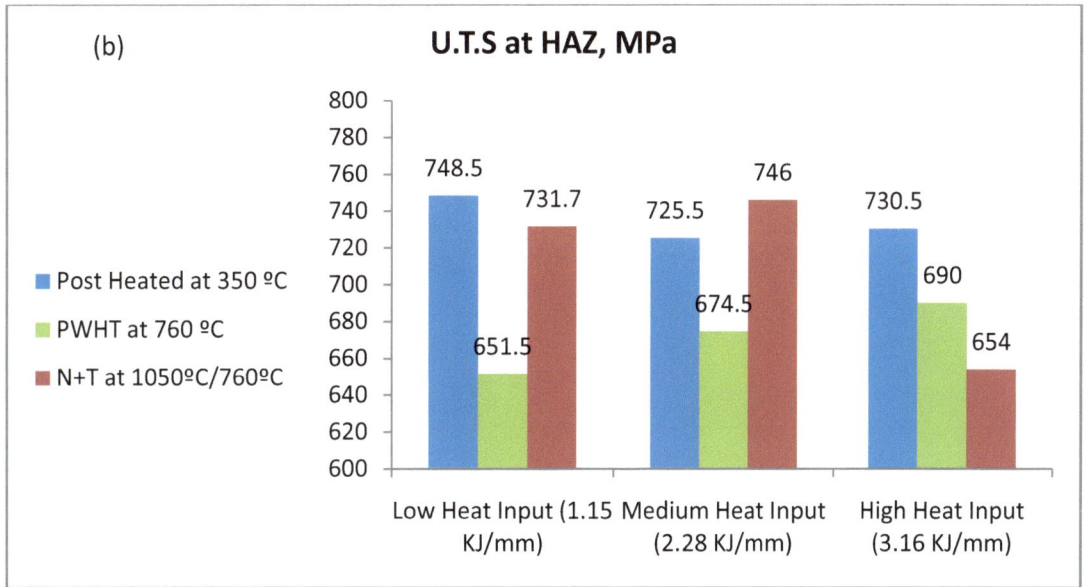

Fig.4.4. Ultimate tensile strength of different heat treatment/heat input conditions. (a) Weld Metal. (b) HAZ.

Though only slight differences appear between the tensile strength values of the HAZ areas of the three as-welded conditions (Fig.4.4 (b)), but, still the lowest heat input condition resulted in the highest UTS value in the as welded condition.

After the normalizing and tempering treatment the results were opposite. The weld metal of high heat input became the strongest; and, the low heat input became the weakest (see Fig 4.4 (a)). However, the difference between UTS of low and medium heat inputs was small. SEM microstructure obtained for the weld metals shows finer grains microstructure associated with the high heat input condition, and nearly the same grain size for both low and medium heat inputs after normalizing treatment, Fig (4.36). It has been shown that a small percentage of delta ferrite is present within the weld metals of different heat inputs after normalizing and tempering treatment and its volume fraction decreases with the increase in heat input (Fig. 4.43 and Para 4.2.2). As for HAZs after tempering at 760 °C, an increase of the tensile strength occurred with the increase in heat input. Low UTS was obtained for the high heat input condition after normalizing and tempering. In fact, the lighter-etching optical microstructure obtained in the HAZ for this condition reflects softer tempered martensite (Low carbon) than that of low and medium heat input conditions (Fig.4.50), and the same was proved true through ferrite content measurements, table(4.10)). This may be also due to the large grain sizes within the HAZ that resulted from the grain coarsening effect at the high heat input **[88]**.

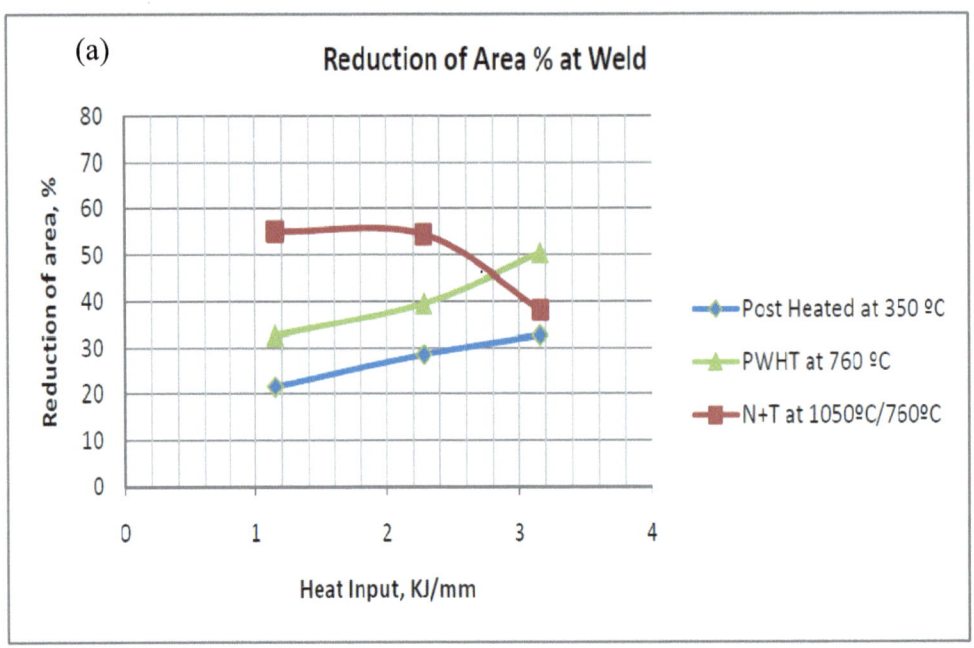

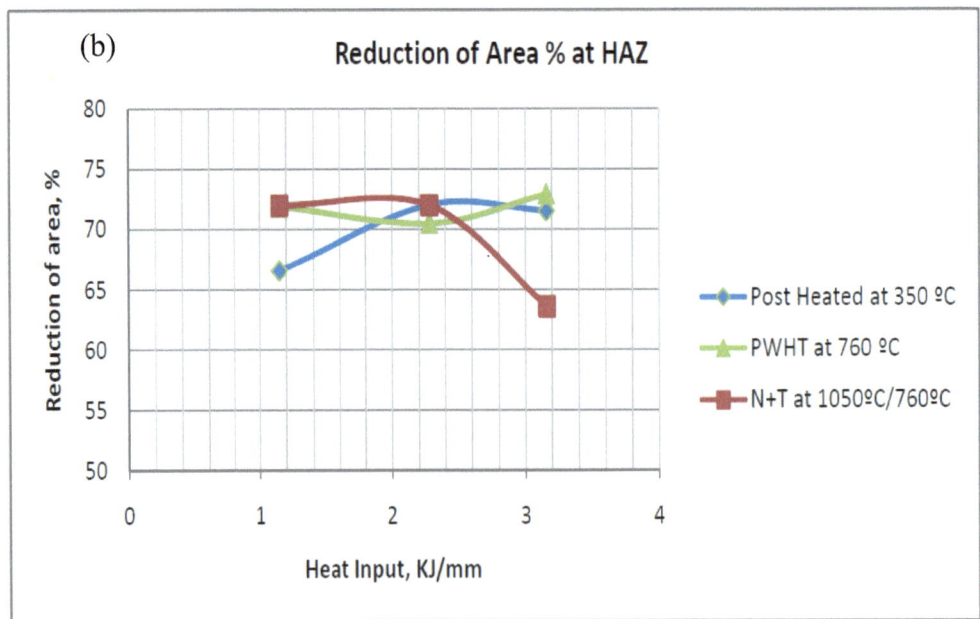

Fig. 4.5. Reduction in area of different heat input/heat treatment conditions. (a) Weld metal. (b) HAZ.

It can be seen from Fig.4.5 (a) that the reduction of area for weld metals of both as welded and subcritical PWHT conditions increases with the increase of heat input. This is matched with decrease of tensile strength with increase of heat inputs for both treatment conditions. The trend changes for the normalized/tempered conditions; the tensile strength increases with the increase in heat input, accompanied with a decrease in reduction in area.

On the other hand, Fig.4.5 (b) shows that the reduction of area at the HAZ of as welded conditions follow the same trend as that of the weld area (Fig. 4.5(a & b)). The reduction in area at HAZ of subcritical PWHT (tempering) did not show considerable change with increasing heat input. The HAZ of normalized/tempered condition increased slightly with increasing heat input.

4. 1.3. Impact test results:

As shown in Fig. 4.6 (a) the impact toughness results agree with the tensile strength and reduction of area for different conditions. The brittle nature of the as welded condition is explained by the large percentage of un-tempered martensite in the microstructure (See Fig.4.28).

The presence of δ-ferrite is another microstructure factor contributing to the impaired toughness of weld. It was reported that steep increase in DBTT occurred with increase in δ-ferrite content in the matrix [89]. δ-ferrite is not preferable in weld as it reduces notch toughness. So, as the low heat input of as welded condition has the highest percentage of un-tempered martensite and δ-ferrite it shows the lowest toughness value. Medium and high heat inputs of the as welded conditions resulted in the same toughness. It is also shown from Fig 4.6 (a) that the subcritical PWHT led to higher impact toughness with narrow differences between the obtained values for different heat inputs. The toughness of the weld metal obtained after normalizing and tempering treatment was much higher than that of the subcritical PWHT for low and medium heat input. Marked decrease in impact toughness occurred after normalizing/tempering treatment of the high heat input weld metal. This is in agreement with the high tensile strength and low reduction of area obtained in specimens welded with high heat input and then subjected to normalizing and tempering treatment (Fig. 4.4(a) and 4.5(a)).

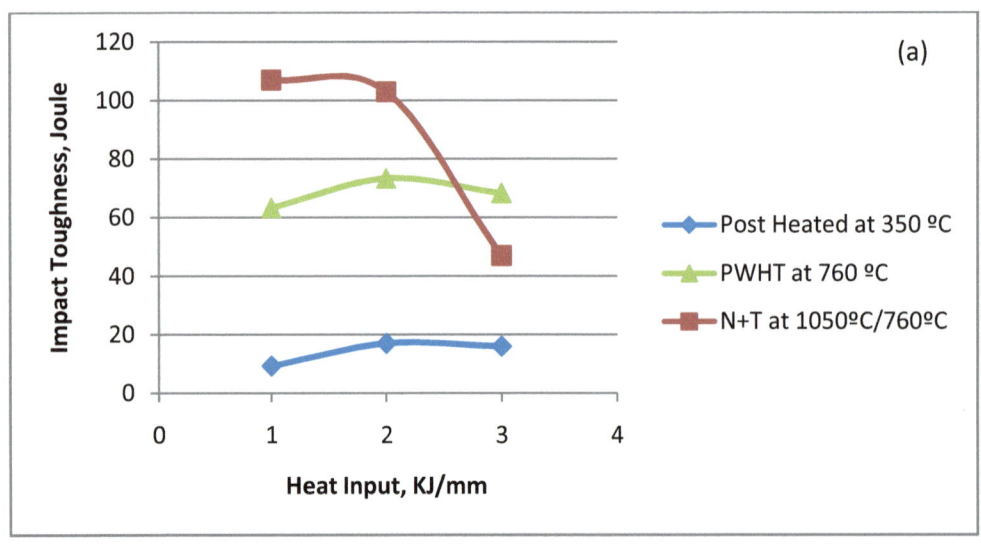

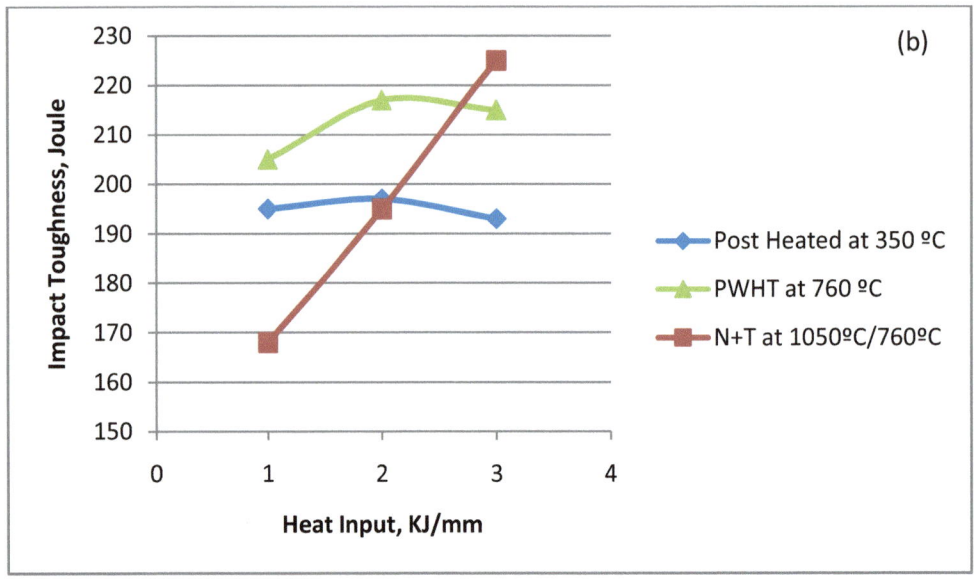

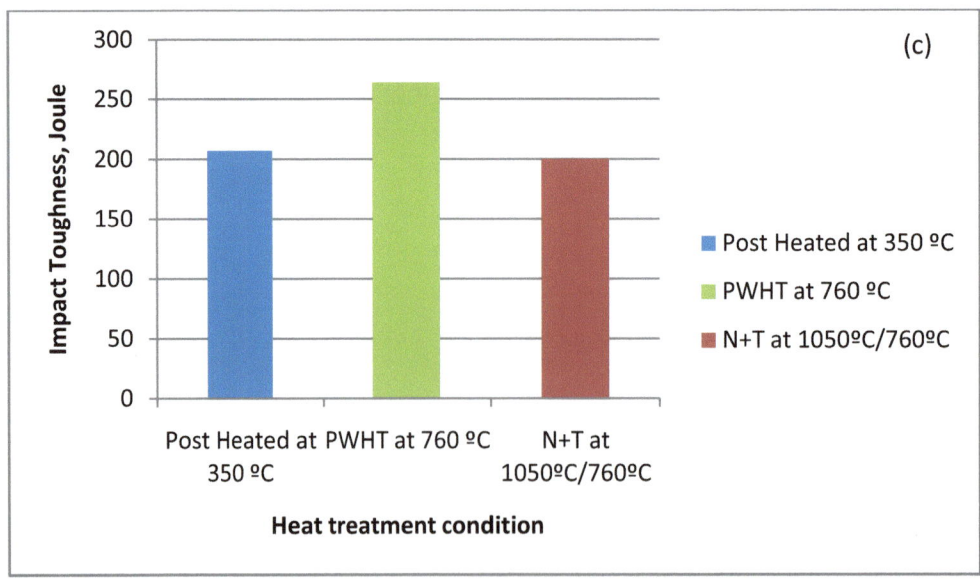

Fig.4.6. Impact toughness of different heat input/heat treatment conditions. (a) Weld metal. (b) HAZ. (c) Base Metal.

The HAZ toughness increases for both subcritical PWHT and normalizing/tempering as shown in Fig. 4.6 (b) with increase in the heat input. However, the subcritical PWHT led to a much higher impact toughness of HAZs at low and medium heat inputs compared to that obtained from normalizing/tempering condition. This trend was reversed at high heat input.

Fig.5.6 (c) shows the effect of the different heat treatments on the impact toughness of the base metal. Re-tempering of the tempered martensite during tempering of the weldment resulted in higher toughness values obtained for the subcritical PWHT condition compared with the as- received condition and the normalized/tempered condition.

4. 1.4. Stress rupture test:

4.1.4.1. Effect of welding and PWHT conditions on time to rupture:

Table 4.1 shows the time to rupture under a stress of 135 MPa at different temperatures for different Heat Treatment/Heat input conditions, and the results are presented in Fig. 4.7, 4.8 and 4.9. From the results obtained at temperature 631°C, which is close to practical applications, it can be seen that normalizing and tempering produced significant increase in the time to rupture mounted to about 90% more than that of the subcritical PWHT (Fig. 4.7). On the other hand, the lower the heat input the higher the time to rupture and this effect becomes more significant in the subcritical PWHT conditions as shown in Fig. 4.7 and 4.9.

Table 4.1: Time to rupture under a stress of 135 MPa at different temperatures for different heat input and heat treatment conditions.

Temp., °C	Condition / Time to Rupture, Hrs					
	L760	M760	H760	L1050	M1050	H1050
631	297	154	156	559	552	496
655	43	69	56	44	107	90
665	28	20	23	56.5	45	40
677	11.2	9.67	7	7.5	16	15

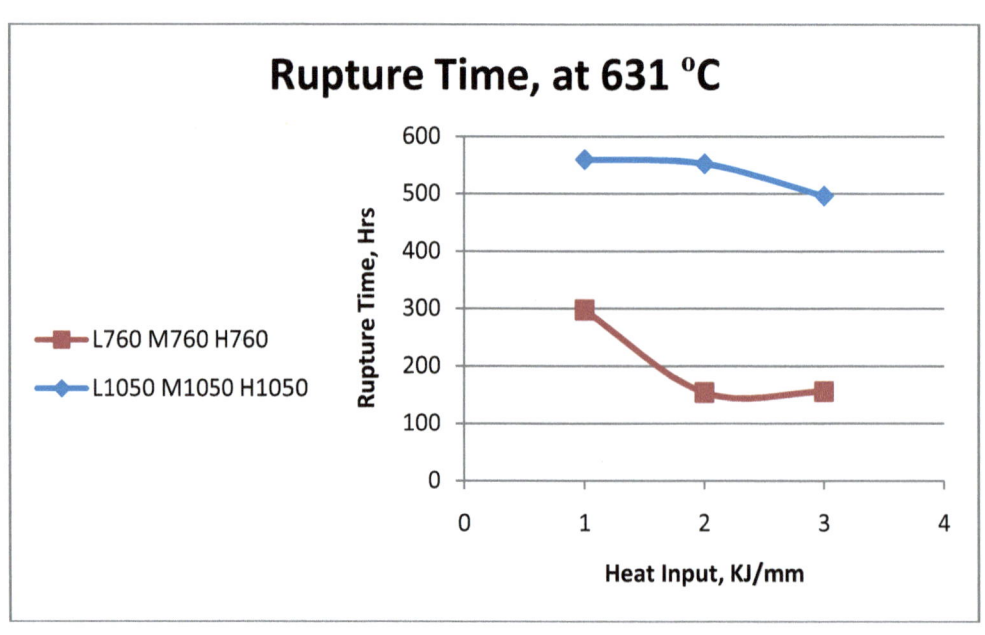

Fig. 4.7 Time to rupture results of different heat input conditions at 135 MPa and Temperature 631 °C.

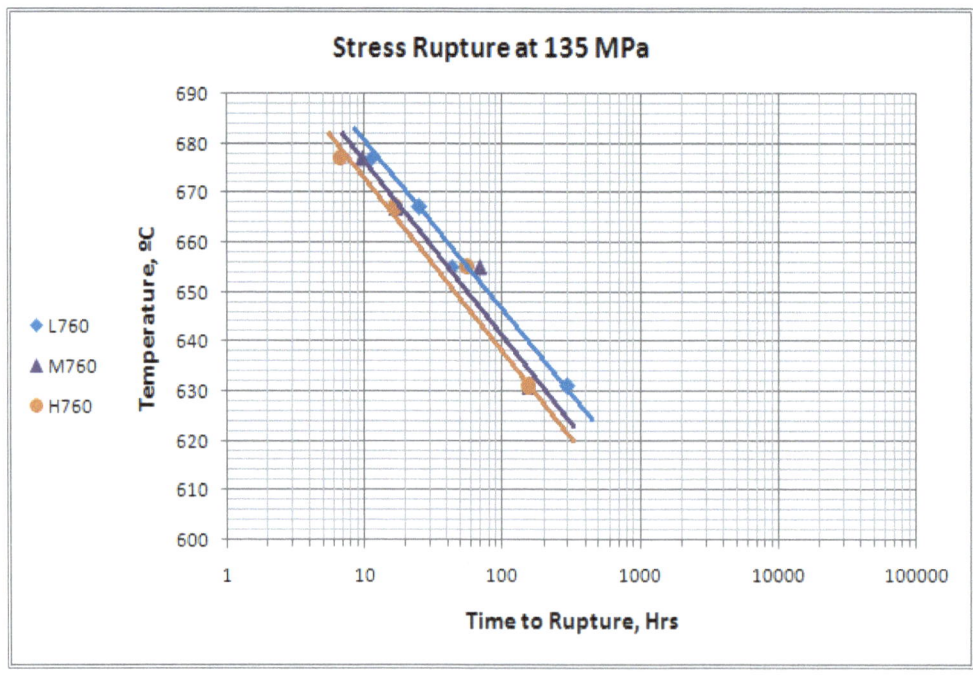

Fig. 4.8 Stress rupture test results of different heat input conditions of Sub-critical PWHT.

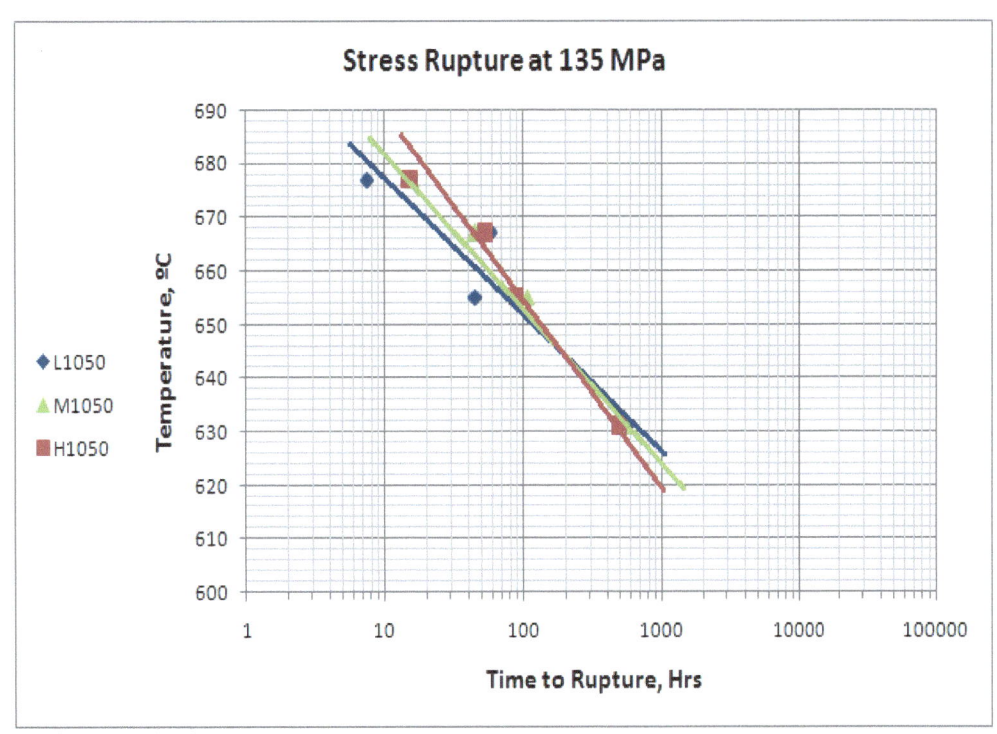

Fig. 4.9. Stress rupture test results of different heat input conditions of Normalized/tempered.

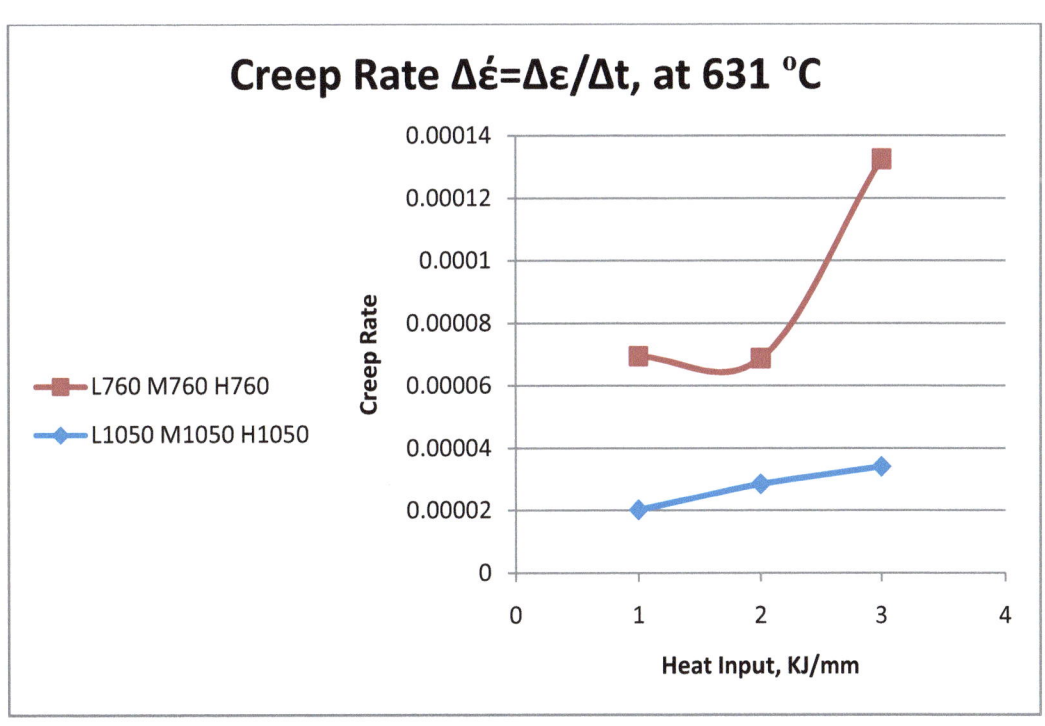

Fig. 4.10. Creep rate resulted from stress rupture test of different heat input and Heat treatment conditions.

Some significant observations were also detected during the analysis of obtained results of stress ruptured specimens. For example the rupture location is affected in by the heat treatment condition and its ability to overcome the softening effect that takes place at the HAZ during welding. The rupture is located at the soft HAZ of all investigated heat input values in the case of subcritical PWHT. The higher the heat input the softer becomes the HAZ and the higher becomes the steady state creep rate as shown in Fig. 4.10. The steady state and the time to rupture become shorter with increasing heat input and increase time to rupture.

On the other hand, the rupture location after normalizing and tempering treatment become different. For low heat input condition the rupture was always located at weld, but for medium and low heat input the rupture location may occur in different areas of the weldment (Weld, HAZ and base). However, at lower temperature (631°C) the rupture was located at HAZ of both medium and high heat input conditions which reflect the inability of this treatment (normalizing plus tempering) to eliminate totally the HAZ softening effect that takes place during welding. Actually, these results are totally matched with the microstructure obtained in weld and HAZ of these conditions. As indicated through Fig (4.49a and 4.50a), the microstructure of weld and HAZ of low heat input condition are identical, and this reflects the high degree of microstructure homogeneity that occur after normalizing followed by tempering , but the rupture is more likely to be located at the weld as a result of the occurrence δ-ferrite in the weld metal zone, Fig. (4.43a). This degree of homogeneity decreases as the heat input increase, and become more significant at high heat input condition, Fig (4.49c and 4.49c), and leads to the change of rupture location to be at HAZ. On the other hand, the time to rupture was found to be nearly the same for both low and medium heat input of the normalized and tempered conditions, Table 4.1 and Fig. 4.10. The differences in the resulted creep rates with increasing heat input for the normalized/tempered conditions become less marked, but still the grater the heat input, the grater the creep rate. Thus the heat input should be as low as possible to decrease creep rate and increase time to rupture.

The obtained values of the reduction of area after stress rupture test are in agreement with the observed softening; the higher the heat input, the higher the reduction of area, Fig. 4.11.

Figures 4.12 and 4.13 show the general trend of increasing elongation and reduction in area with increasing temperature of stress rupture test.

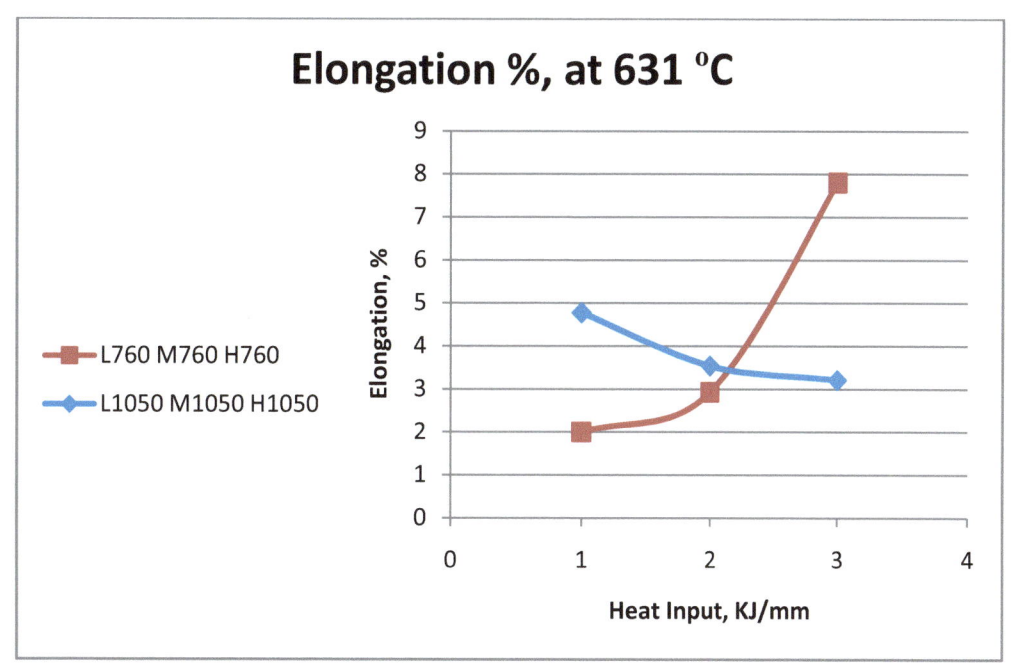

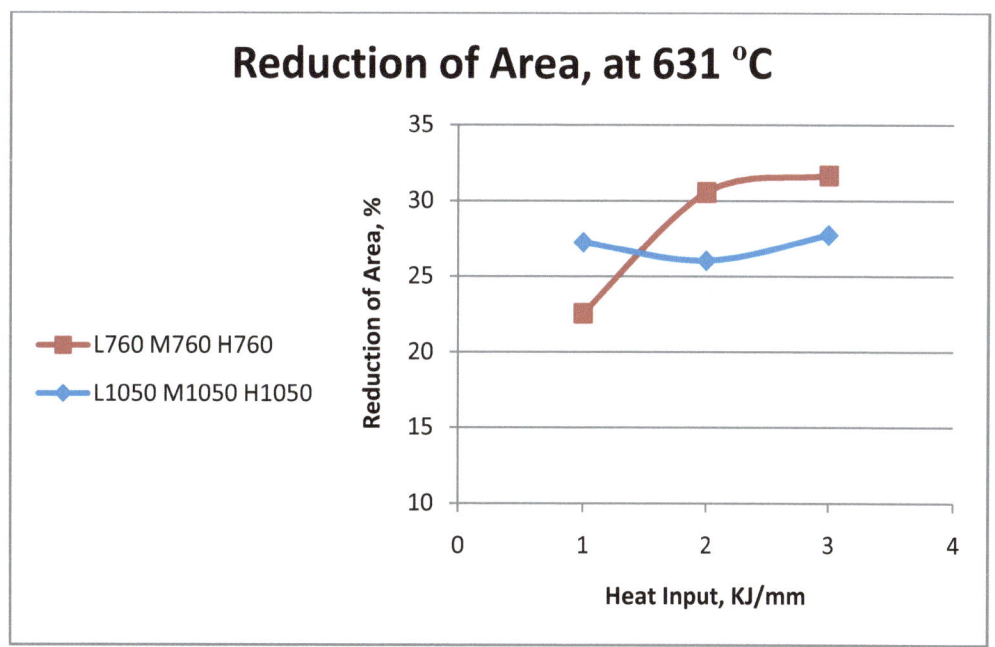

Fig. 4.11 Elongation and Reduction of Area obtained from Stress rupture tests.

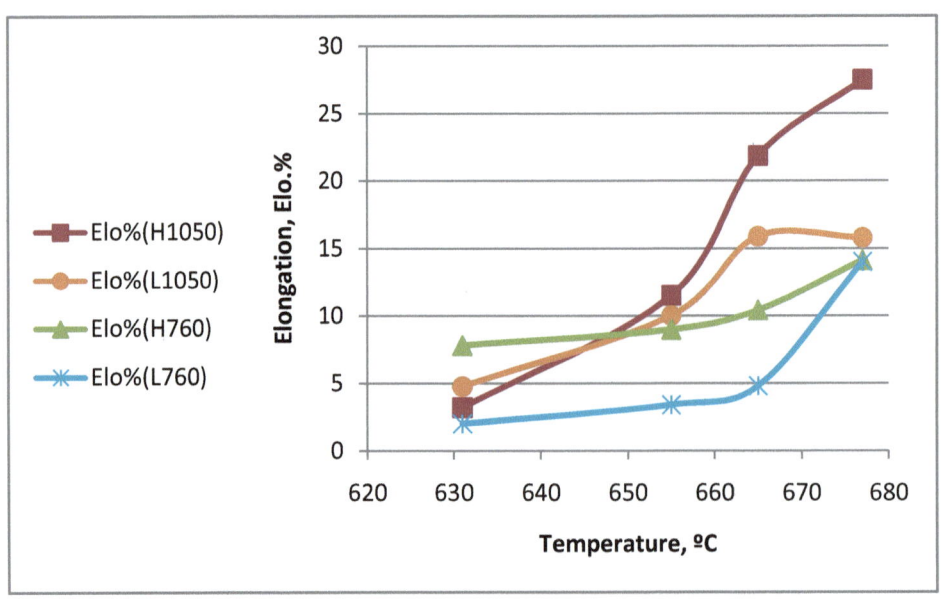

Fig. 4.12 : Elongation of Low and High heat input in both sub-critical and normalizing/tempering PWHT conditions

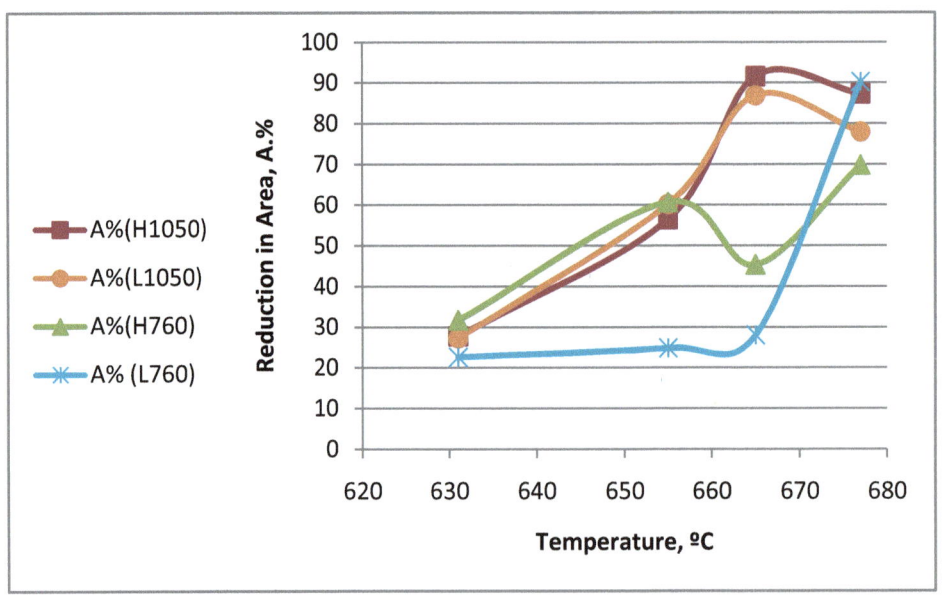

Fig. 4.13 : Reduction in area of Low and High heat input in both sub-critical and normalizing/tempering PWHT conditions

The location of fracture, time to rupture, elongation, and reduction in area are given in Tables from 4.2 to 4.7 and figures from 4.14 to 4.19

Table 4. 2: Stress rupture test results for Low heat input condition followed by PWHT at 760 °C for 3.5 hrs (L760).

Test Temp., °C	Rupture Time (tr), hrs	Minimum Creep Rate ($\dot{\varepsilon}$ min), h^{-1}	Elongation, %	Reduction of Area, %	Fracture Location
677	11.13	------	14.00	90.16	HAZ
665	28	------	4.80	27.96	HAZ
655	43.2	25.32×10^{-5}	3.40	24.77	HAZ
631	297.46	6.93×10^{-5}	2.00	22.56	HAZ

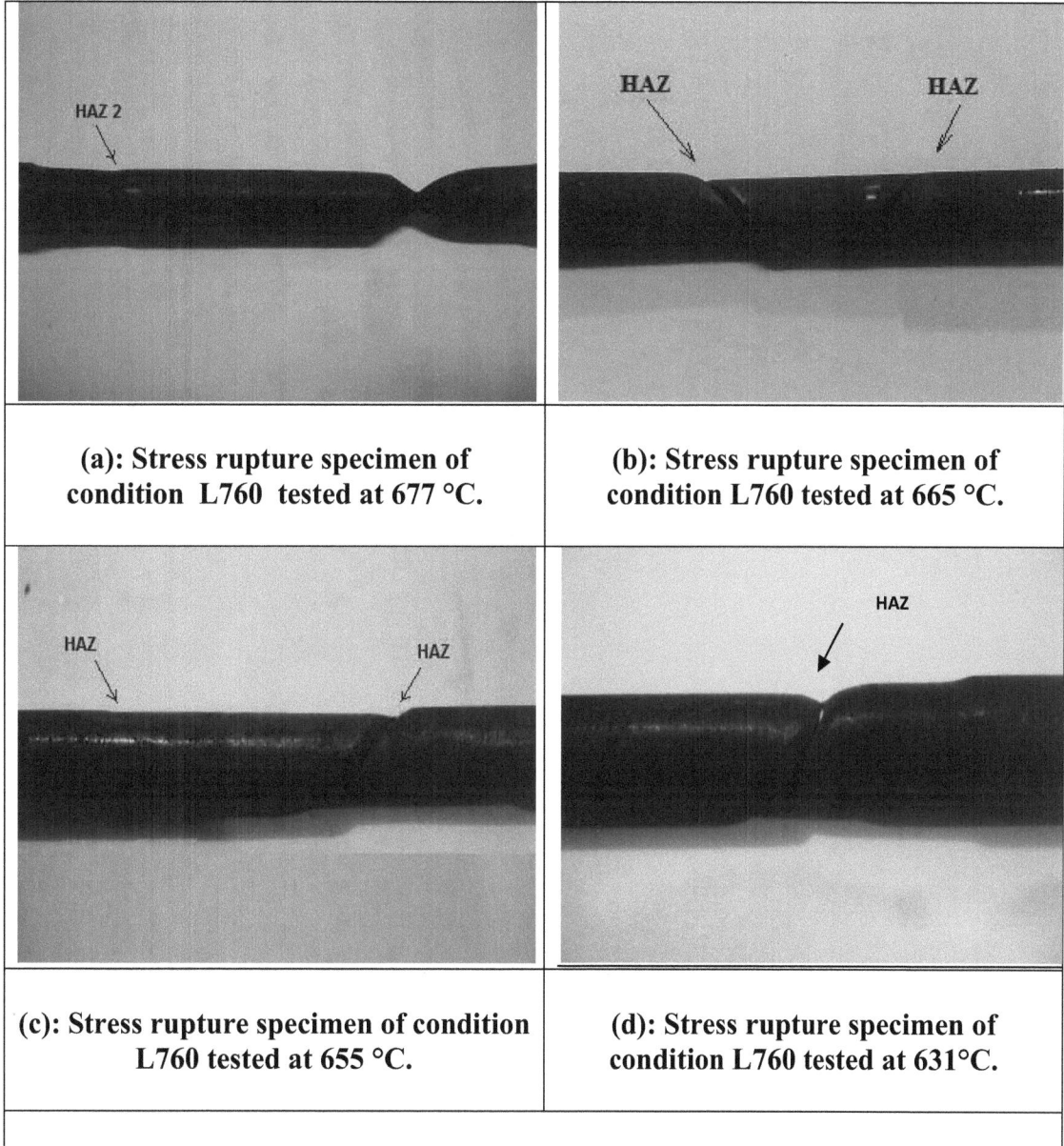

(a): Stress rupture specimen of condition L760 tested at 677 °C.

(b): Stress rupture specimen of condition L760 tested at 665 °C.

(c): Stress rupture specimen of condition L760 tested at 655 °C.

(d): Stress rupture specimen of condition L760 tested at 631°C.

Fig 4.14: Fracture appearance and location for Low heat input condition followed by PWHT at 760 °C for 3.5 hrs (L760).

Table 4.3: Stress rupture test results for Medium heat input condition followed by PWHT at 760 °C for 3.5 hrs (M760).

Test Temp., °C	Rupture Time (tr), hrs	Minimum Creep Rate ($\dot{\varepsilon}$ min), h^{-1}	Elongation, %	Reduction of Area, %	Fracture Location
677	9.67	------	15.07	83.04	HAZ
665	20.13	------	6.36	56.44	HAZ
655	68.75	59.82x10^{-5}	10.34	57.69	HAZ
631	155	6.874 x 10^{-5}	2.93	30.56	HAZ

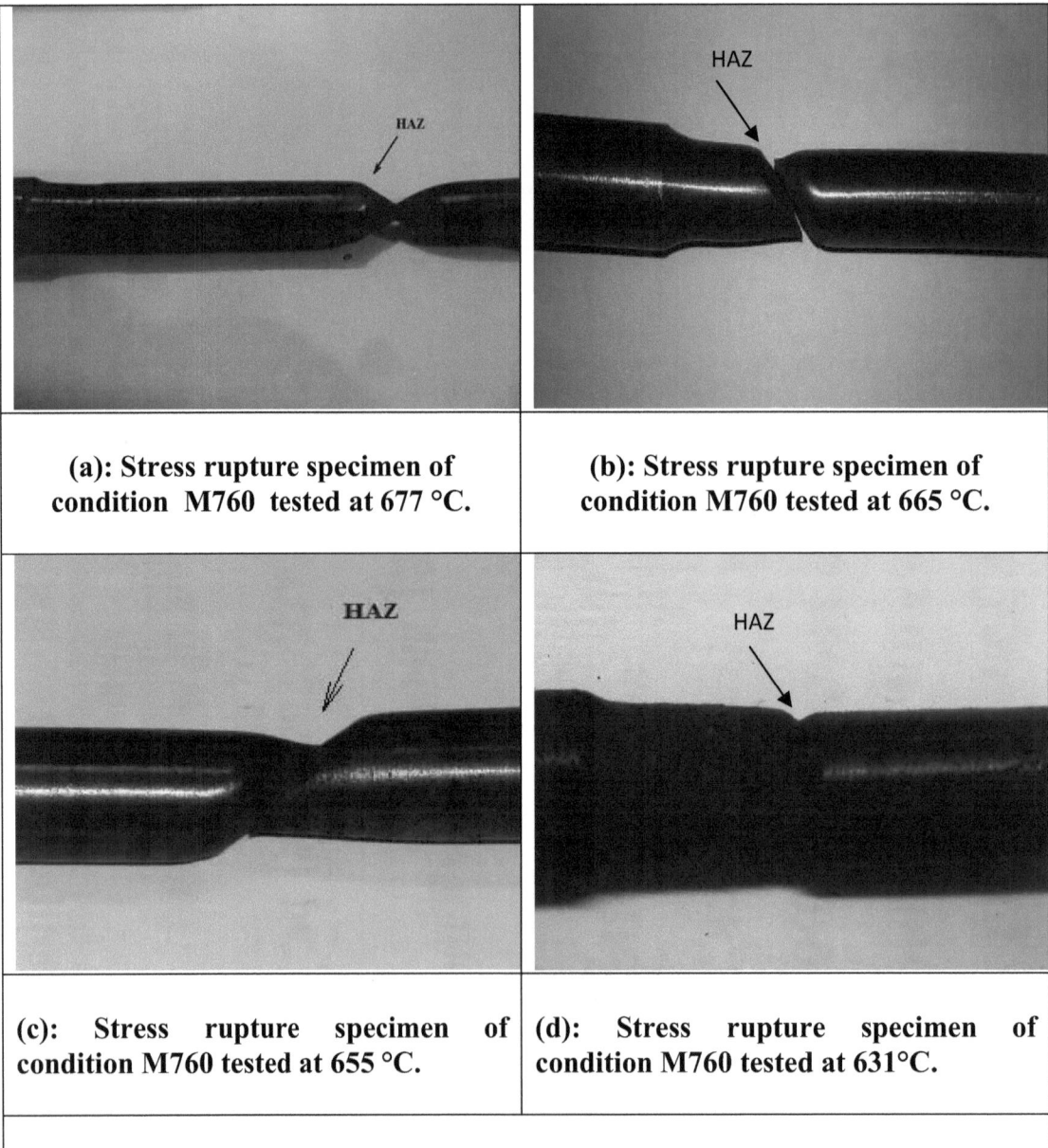

(a): Stress rupture specimen of condition M760 tested at 677 °C.

(b): Stress rupture specimen of condition M760 tested at 665 °C.

(c): Stress rupture specimen of condition M760 tested at 655 °C.

(d): Stress rupture specimen of condition M760 tested at 631°C.

Fig 4.15: Fracture appearance and location for medium heat input condition followed by PWHT at 760 °C for 3.5 hrs (M760).

Table 4.4: Stress rupture test results for High heat input condition followed by PWHT at 760 °C for 3.5 hrs (H760):

Test Temp., °C	Rupture Time (tr), hrs	Minimum Creep Rate ($\dot{\varepsilon}$ min), h^{-1}	Elongation, %	Reduction of Area, %	Fracture Location
677	6.78	------	14.13	69.75	HAZ
665	23.2	------	10.40	45.24	HAZ
655	55.5	59.54x10^{-5}	8.97	60.63	HAZ
631	155.75	13.25 x 10^{-5}	7.80	31.68	HAZ

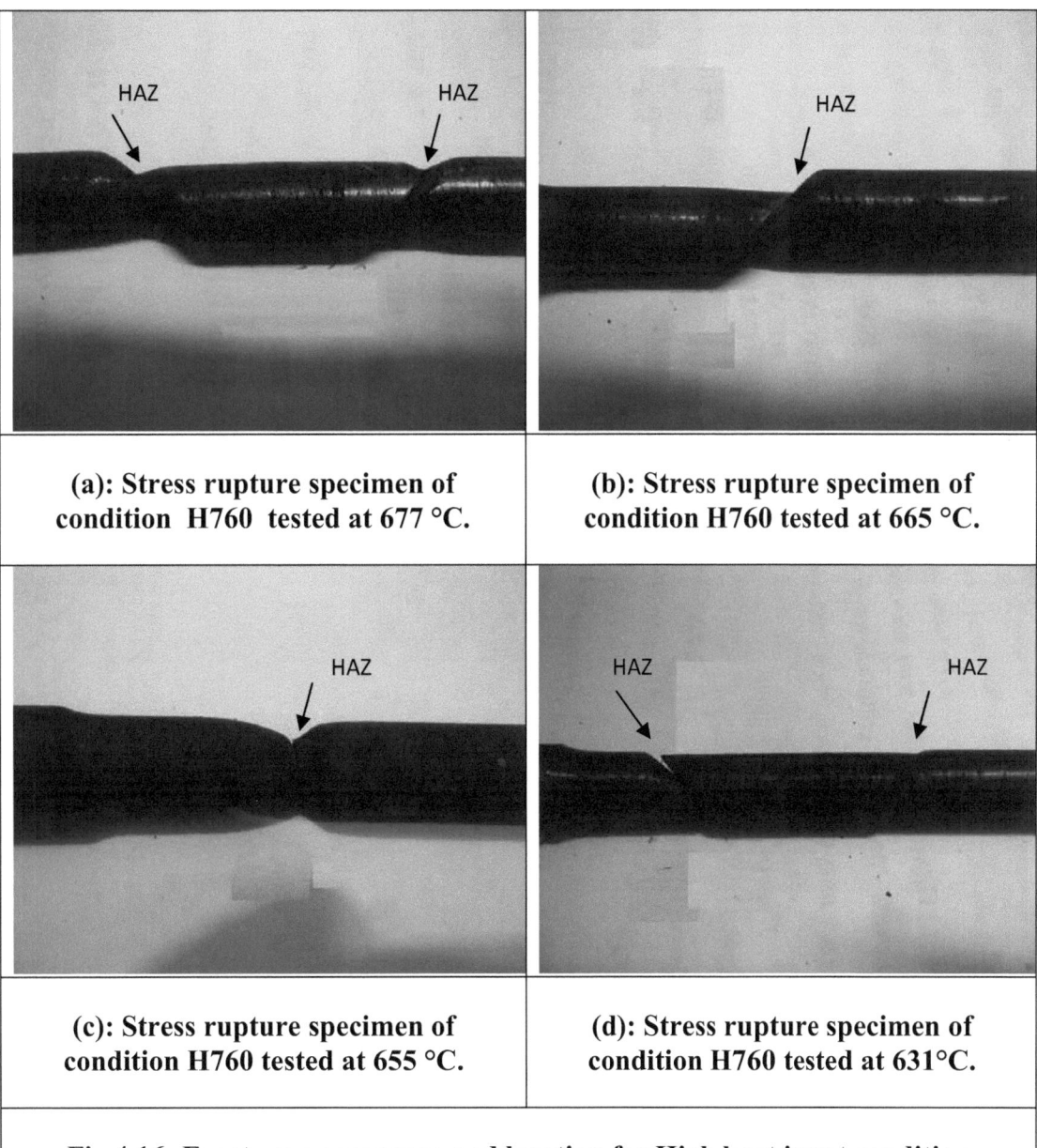

(a): Stress rupture specimen of condition H760 tested at 677 °C.

(b): Stress rupture specimen of condition H760 tested at 665 °C.

(c): Stress rupture specimen of condition H760 tested at 655 °C.

(d): Stress rupture specimen of condition H760 tested at 631°C.

Fig 4.16: Fracture appearance and location for High heat input condition followed by PWHT at 760 °C for 3.5 hrs (H760).

Table 4.5: Stress rupture test results for Low heat input condition followed by normalizing at 1050 °C for 0.5 hr and tempering at 760 °C for 3.5 hrs (L1050):

Test Temp., °C	Rupture Time (tr), hrs	Minimum Creep Rate ($\dot{\varepsilon}$ min), h^{-1}	Elongation, %	Reduction of Area, %	Fracture Location
677	7.5	-----	15.77	77.91	Weld
665	57.5	-----	15.87	86.84	Weld
655	44	47.92×10^{-5}	10.00	60.18	Weld
631	559.33	2.017×10^{-5}	4.77	27.25	Weld

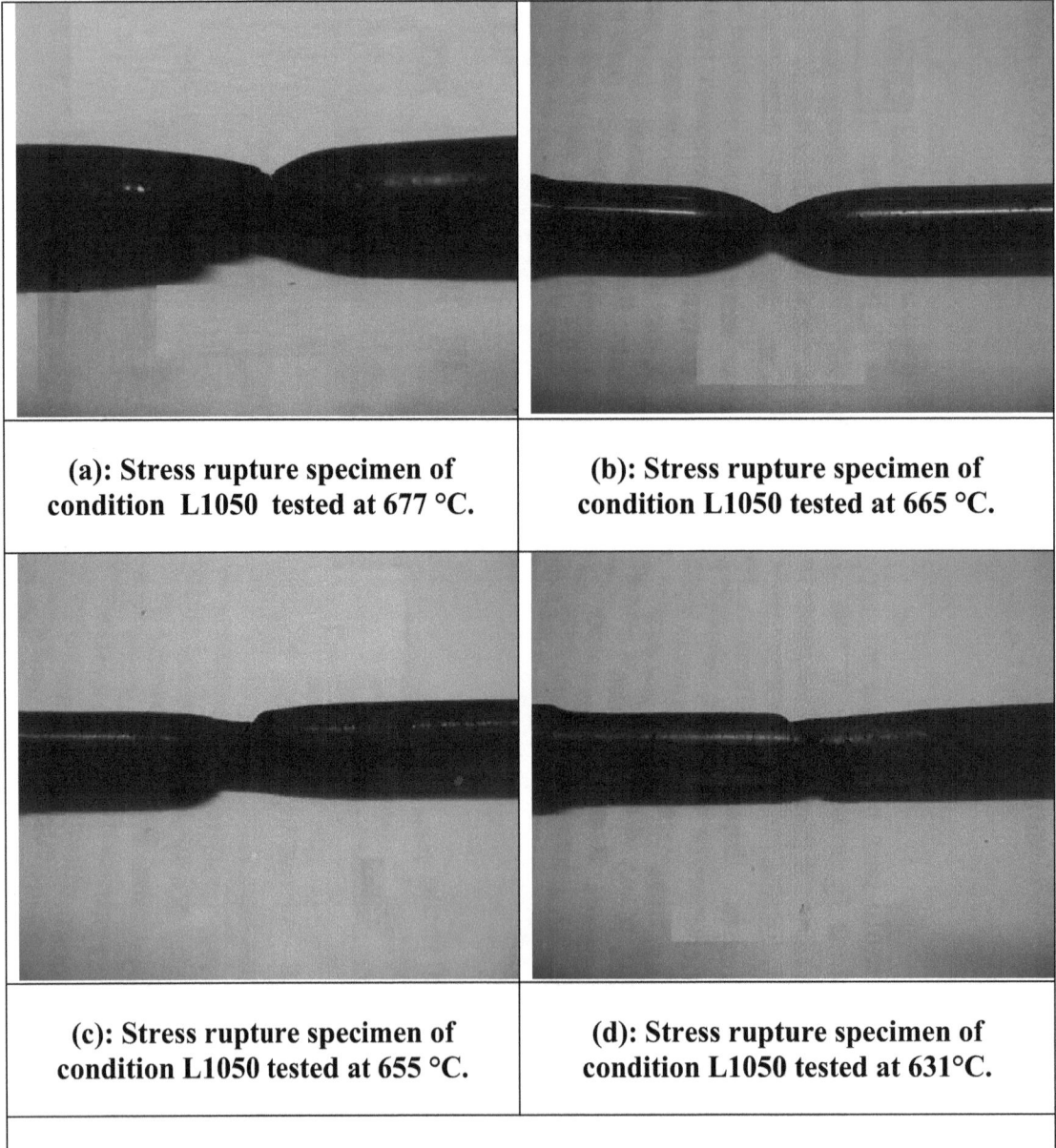

(a): Stress rupture specimen of condition L1050 tested at 677 °C.

(b): Stress rupture specimen of condition L1050 tested at 665 °C.

(c): Stress rupture specimen of condition L1050 tested at 655 °C.

(d): Stress rupture specimen of condition L1050 tested at 631°C.

Fig 4.17: Fracture appearance and location for Low heat input condition followed by normalizing at 1050 °C for 0.5 hr and tempering at 760 °C for 3.5 hrs (L1050).

Table 4.6: Stress rupture test results for Medium heat input condition followed by normalizing at 1050 °C for 0.5 hr and tempering at 760 °C for 3.5 hrs (M1050):

Test Temp., °C	Rupture Time (tr), hrs	Minimum Creep Rate ($\dot{\varepsilon}$ min), h^{-1}	Elongation, %	Reduction of Area, %	Fracture Location
677	15.57	-----	16.84	86.51	Weld
665	45	-----	14.80	80.64	Base Metal
655	107.26	28.44x10^{-5}	17.72	92.96	Base Metal
631	552	2.85x10^{-5}	3.54	26.04	HAZ

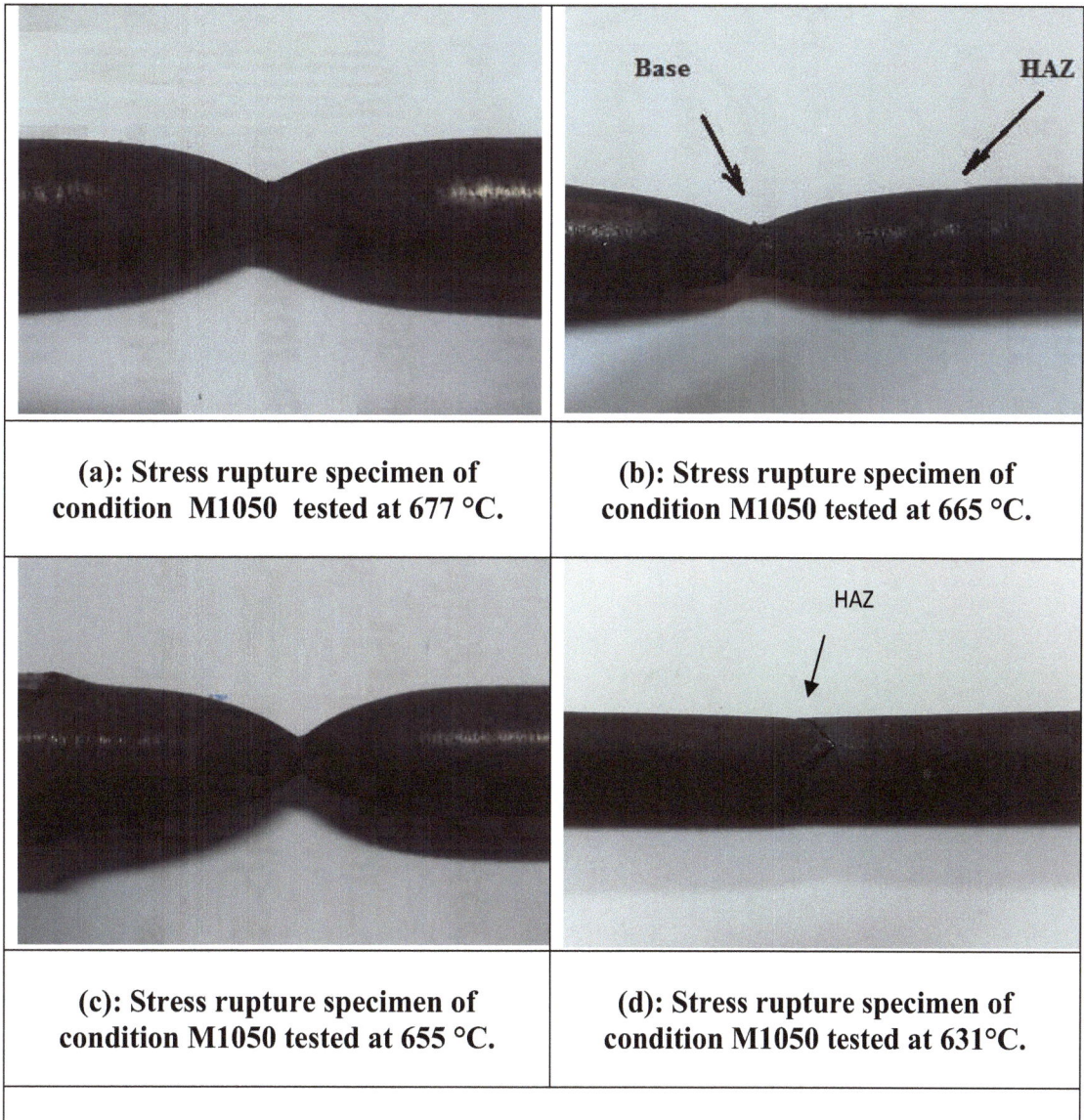

(a): Stress rupture specimen of condition M1050 tested at 677 °C.

(b): Stress rupture specimen of condition M1050 tested at 665 °C.

(c): Stress rupture specimen of condition M1050 tested at 655 °C.

(d): Stress rupture specimen of condition M1050 tested at 631°C.

Fig 4.18: Fracture appearance and location for medium heat input condition followed by normalizing at 1050 °C for 0.5 hr and tempering at 760 °C for 3.5 hrs (M1050)

Table 4.7: Stress rupture test results for High heat input condition followed by normalizing at 1050 °C for 0.5 hr and tempering at 760 °C for 3.5 hrs (H1050):

Test Temp., °C	Rupture Time (tr), hrs	Minimum Creep Rate ($\dot{\varepsilon}$ min), h^{-1}	Elongation, %	Reduction of Area, %	Fracture Location
677	15.16	-----	27.50	87.24	Weld
665	53.5	-----	21.85	91.52	Base Metal
655	89.75	46.13x10^{-5}	11.51	56.44	Weld
631	495	3.41x10^{-5}	3.21	27.75	HAZ

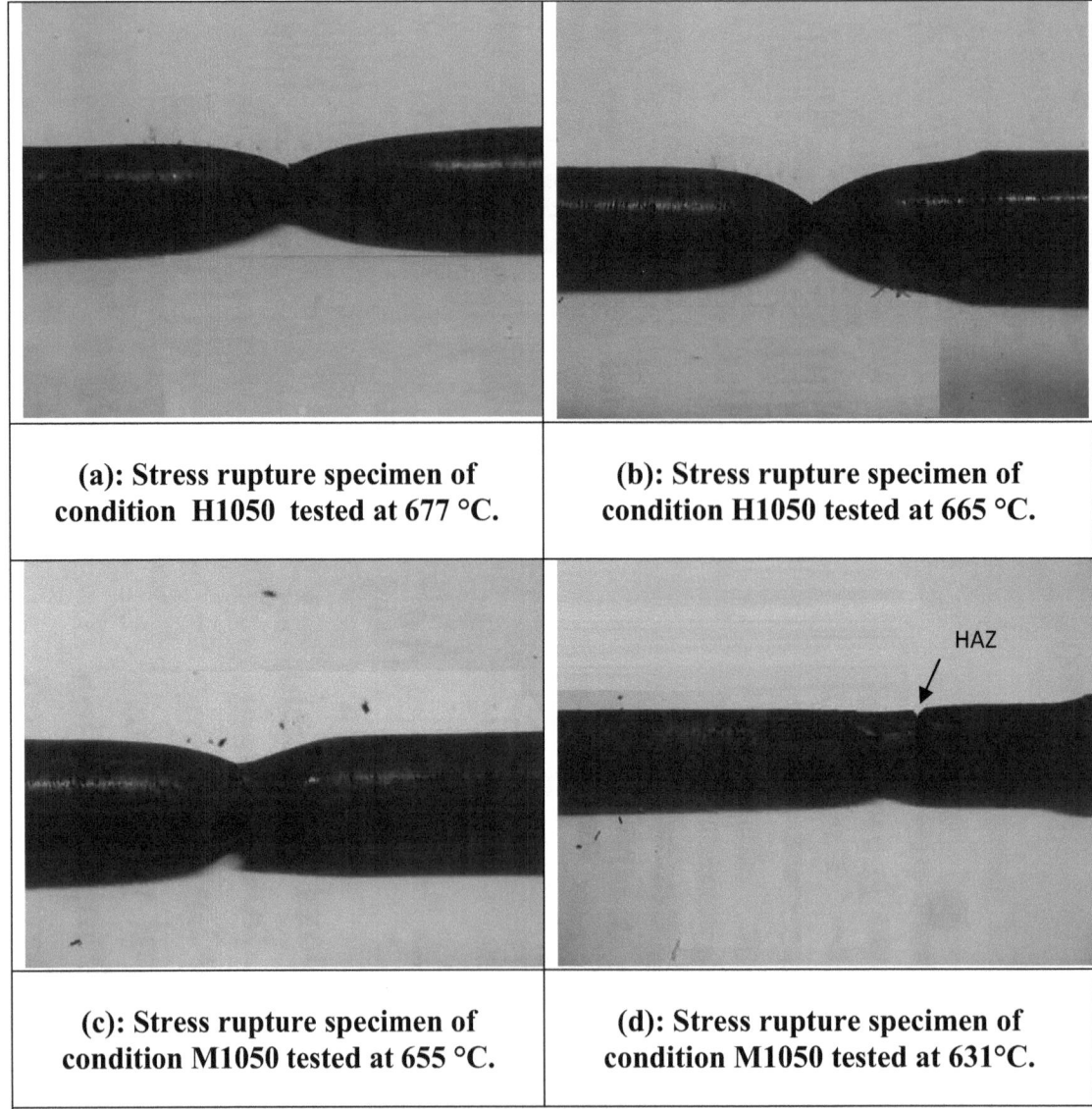

(a): Stress rupture specimen of condition H1050 tested at 677 °C.

(b): Stress rupture specimen of condition H1050 tested at 665 °C.

(c): Stress rupture specimen of condition M1050 tested at 655 °C.

(d): Stress rupture specimen of condition M1050 tested at 631°C.

Fig 4.19: Fracture appearance and location for High heat input condition followed by normalizing at 1050 °C for 0.5 hr and tempering at 760 °C for 3.5 hrs (H1050)

4.1.5. Stress corrosion cracking results:

No cracks were observed. Pitting increased in number and severity with increasing heat input during welding as described in the figures from 4.20 to 4.23.

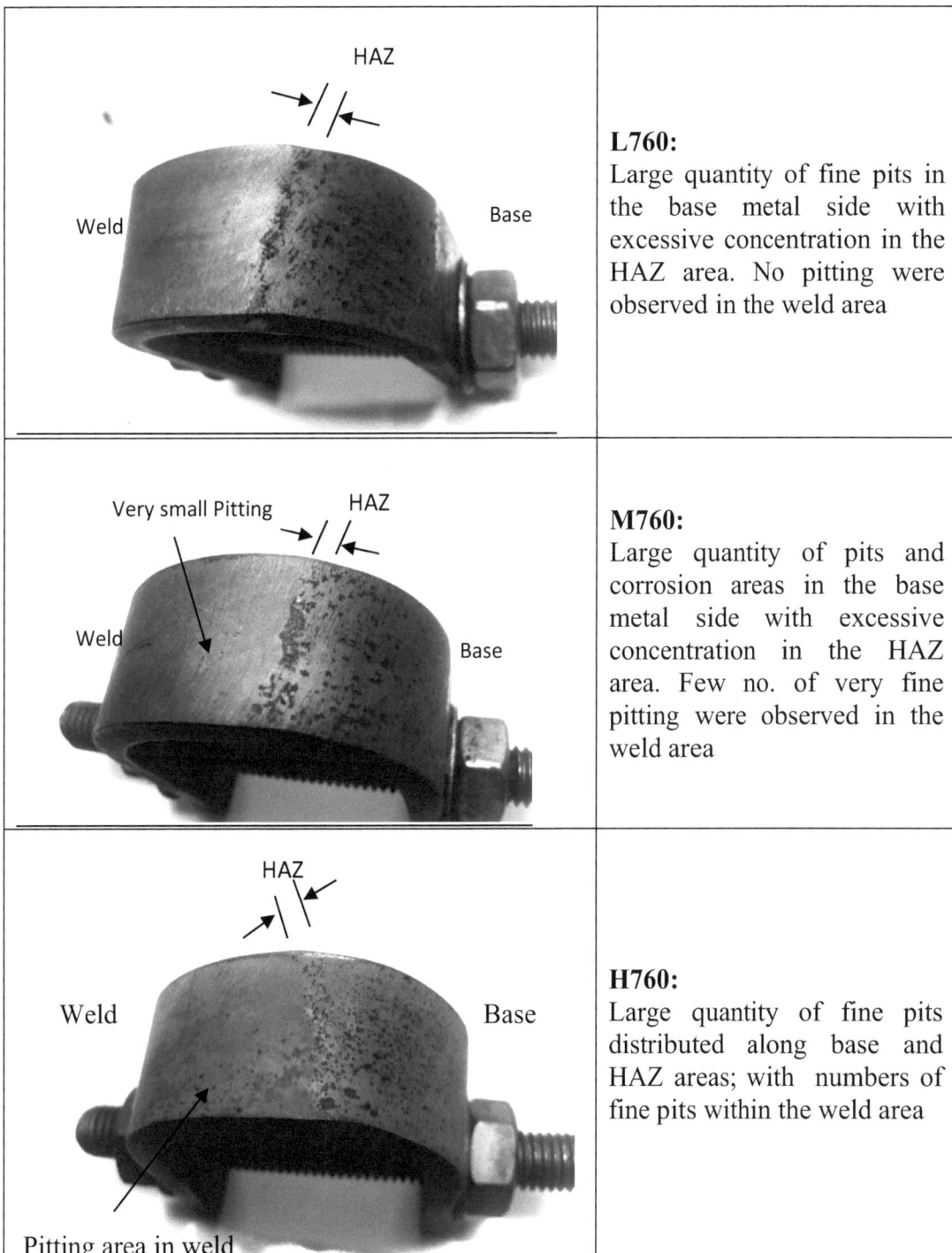

Fig. 4.20: Weld area of L760, M760 and H760 tested for 0.5 day. Percentage of pitting in the weld area increased with increase in the heat input during welding for the tempered conditions. Low heat input condition showed no pitting.

	**L1050**: Large quantity of pits and corrosion areas in the base metal side starting from an area near the HAZ/Base line. No pitting were observed in the weld area.
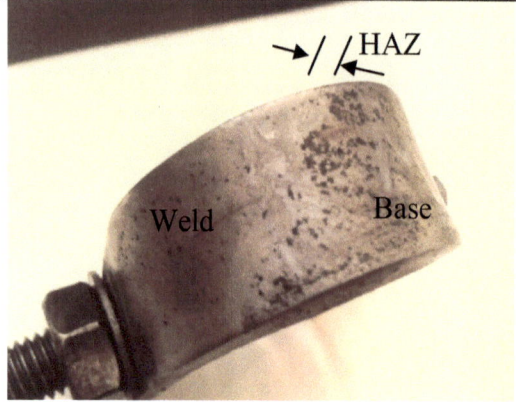	**M1050**: Large quantity of pits and corrosion areas in the base metal side starting from the HAZ/Base line with d pitting in the weld area.
	**H1050**: Large quantity of fine pits in the base metal side starting from the HAZ/Base line with large numbers of fine pitting within the weld area.

Fig. 4.21: Weld area of H1050, M 1050 and L1050 tested for 0.5 day. Percentage of pitting in the weld area increased with increase in the heat input during welding for the normalized/tempered conditions. Low heat input condition showed no pitting.

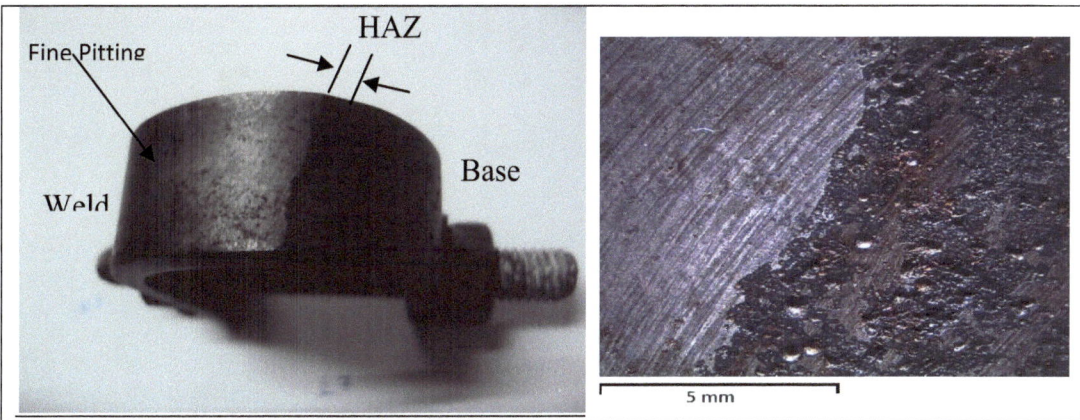

L760: Number of pits and corroded areas increased in the base metal side with excess concentration in the HAZ area. Fine pitting was observed in the weld area tested for 5 days.

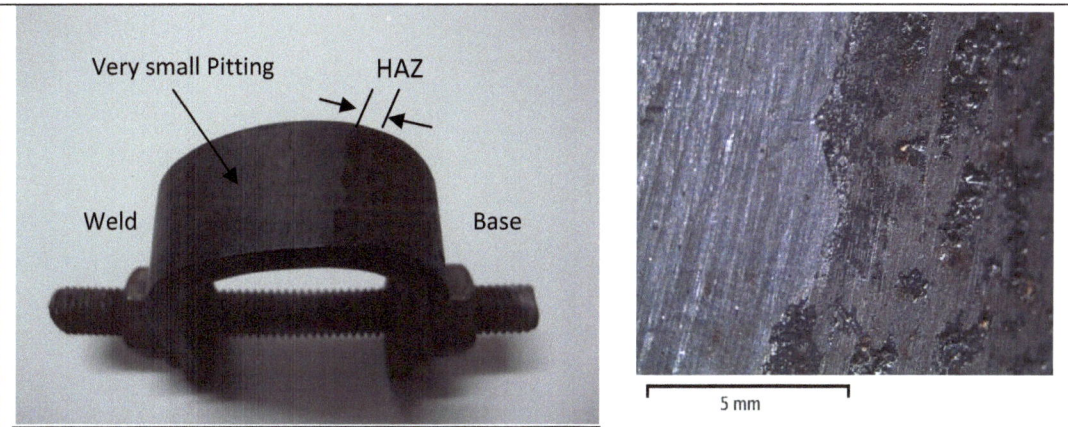

M760: Number of pits and corroded areas increased in the base metal side with excess concentration in the HAZ area. The percentage of fine pitting was increased in the weld area tested for 7.5 day.

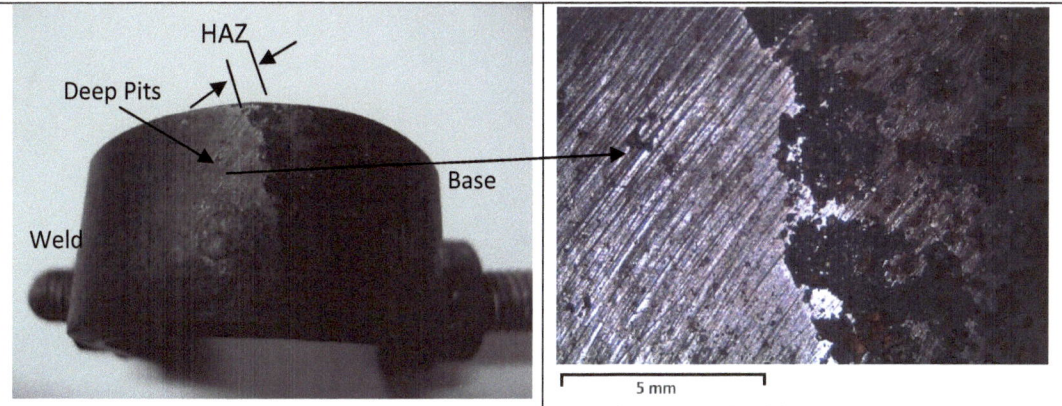

H760: Number of pits and corroded areas increased severely along base and HAZ areas; with numbers of fine distributed pitting within the weld area tested for 4 day.

Fig 4.22: Weld area of L760, M760 and H760 tested for 5, 7.5 and 4 days respectively

L1050: Large spots of corroded areas at the HAZ/Base boundary. No pitting was observed in the weld area of condition L1050 tested for 7 day.

M1050: No. of pits and corroded areas increased and concentrated in the HAZ and on the other hand distributed pits with small and deep corrosion areas in the weld area of condition M1050 tested for 4 day.

H1050: Number of pits with sever corroded areas in the HAZ increased with increase in numbers of fine distributed pits within the weld area of condition H1050 tested for 5.5 day.

Fig. 4.23: Weld area of H1050, M 1050 and L1050 tested for 7, 4 and 5.5 days respectively. Percentage of corrosion in the weld and HAZ areas increased with increase in the heat input during welding for the normalized/tempered conditions.

The stress corrosion cracking tests conducted on all conditions illustrates the good resistance of this material and its weldments to the stress corrosion attack whatever the heat input or heat treatment is; but on the same time give an indication on the degree of pitting resistance related to the used heat input and heat treatment.

Fig. 4.20 and 4.22, for subcritical PWHT conditions, indicate the decrease in pitting resistance with the increase of heat input for weld metal and HAZ area, but for HAZ the pitting resistance is lower than weld metal and pitting resistance of the base metal is the lowest. On the other hand, and for normalized tempered conditions, Fig. 4.21 and 4.23, indicates the same results indicated for subcritical PWHT conditions but reflects better resistance than that of sub-critical PWHT. However, the microstructure investigation reflects the relation between the measured ferrite content and pitting resistance, the higher the ferrite content, the lower the pitting resistance (The softer the tempered martensite, the lower the pitting resistance).

However, it is clear from this study that the corrosion behavior will have a direct effect on the creep life of the weldment and this is affected by the used heat input and heat treatment condition.

4. 2. Microstructure:

4. 2. 1. Base metal

In the as-received, sub-critical PWHT and normalized /tempered conditions, the microstructure of the P91 base metal consists of fully tempered martensite (Fig. 4.24(a-c)). The increase in the measured ferrite volume fraction is used as indication for the degree of decomposition of martensite during sub-critical PWHT, Fig.4.24(b). Small precipitate particles could be identified on the prior austenite grain boundaries and within the grains, (Fig. 4.24(a-c)). SEM investigation conducted on the as-received base metal (Fig. 4.24(a)) shows the lath martensitic structure, as well as the precipitate particles embedded along prior austenite grains and the lath boundaries; these precipitates are possibly the $M_{23}C_6$ carbides as identified in earlier studies[90, 91, 92]. Much finer precipitate particles are found both along the boundaries and within the laths; these particles are most probably of niobium-rich or vanadium-rich carbonitrides and carbides of type MC, M_2C as has been reported in Grade 91 steels [93]. EDS analysis (Fig.4.25 and table 4.8) conducted on the as-received base metal confirmed the presence of $M_{23}C_6$ (M=Cr, Fe, Mo) precipitates across different areas within the tempered martensite. X-ray diffraction patterns of the three base metal conditions indicated that the structure consisted mainly of martensitic matrix without any retained austenite (Fig. 4.26 and table 4.9). Also, no delta ferrite was detected in the microstructure.

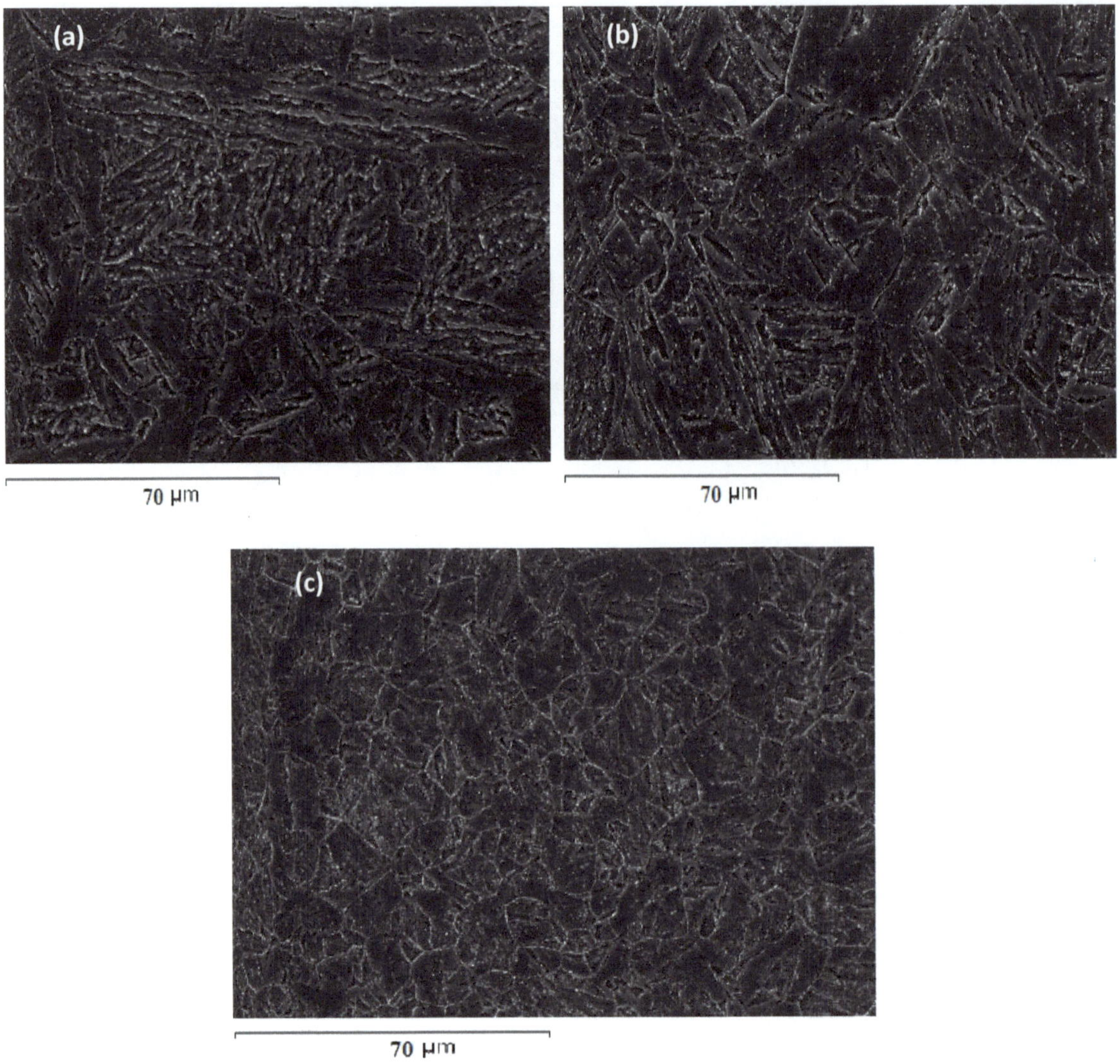

Fig. 4.24. SEM Microstructure of P91 Base Metal:
(a) As Received Base Metal, 91 % measured ferrite (F). (b) Sub-critical PWHT
Base Metal, 95% F. (c) Normalized/Tempered Base Metal, 93% F.

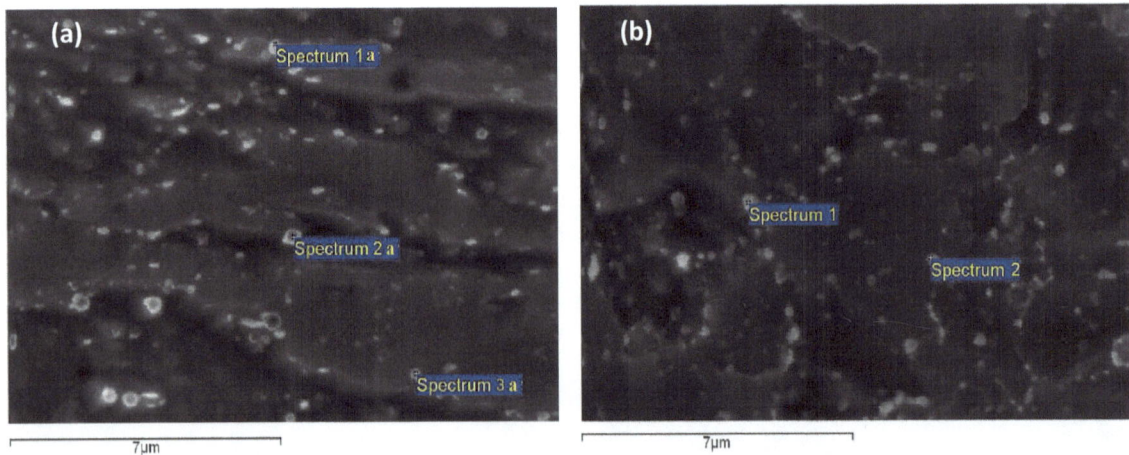

Fig. 4.25. Location of precipitates subjected to EDS in the As received Base Metal.
(a) Along Martensite lathes. (b) Along Ferrite grains.

Table 4.8: Results of EDS analysis related to precipitates shown in Fig.4.25, wt %.

Element	C	Si	V	Cr	Mn	Fe	Ni	Nb	Mo
Spec.1a	42.98	----	0.2	6.13	0.39	49.23	0.29	----	0.79
Spec.2a	39.02	0.73	0.26	5.43	0.77	52.37	0.25	0.23	0.95
Spec.3a	41.10	0.23	0.09	5.20	0.38	52.27	0.24	0.05	0.43
Spec.1	40.30	0.17	0.15	7.72	----	50.56	----	0.05	1.05
Spec.2	42.10	0.21	0.16	5.17	0.23	51.83	----	----	0.28

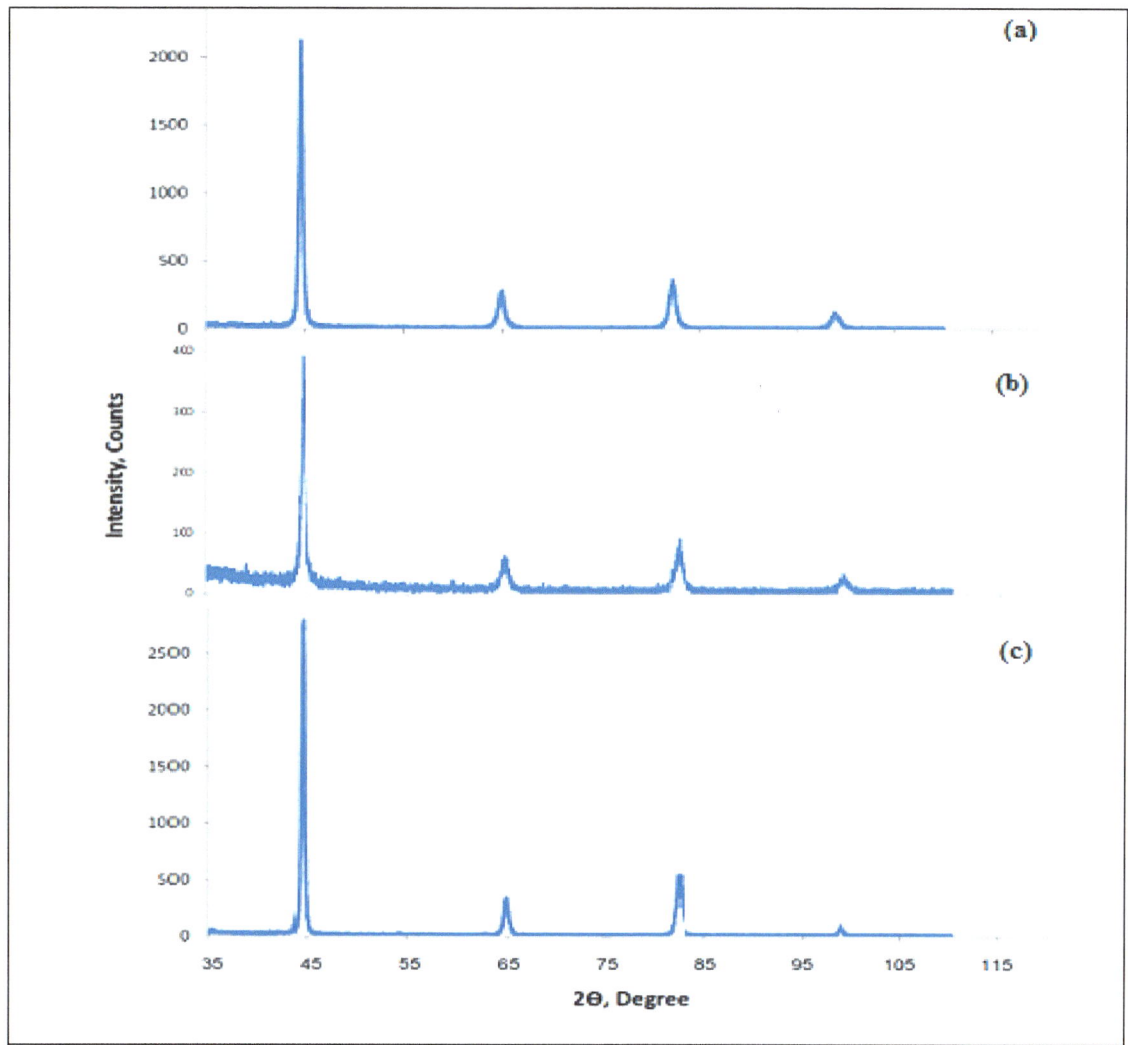

Fig. 4.26: X-Ray diffraction pattern of P91 base metal after different treatment conditions: (a) B350. (b) B760. (c) B1050.

Table 4.9: Peaks analysis of X-Ray diffraction pattern of P91 base metal after different treatment conditions.

$2\Theta^\circ$	44.52	64.90	82.20	98.76
d-Spacing, Å	2.03484	1.43665	1.17266	1.01478
Matched with Phase/(*hkl*)	Martensite (110)	Martensite (200)	Martensite (211)	Martensite (220)

4. 2. 2. Weld metal and HAZ:

Fig. 4.27 illustrates the expected microstructure of the weld metal and HAZ as a function of the peak temperature gradient along the P91 weldement [94] and also the CCT diagram of steel P91. From Fig 4.27 (CCT diagram) it is seen that the likelihood of forming ferrite during solidification from welding temperature is negligible due to the required very slow cooling rate. The optical micrographs of the three welding heat inputs of as-welded conditions are shown through Fig. 4.28, which show a typical un-tempered martensite structure.

Different microstructures are observed in the weld metal due to the partial reheating action caused by each pass on the previous one (see Fig. 4.30). SEM microstructure conducted for as-welded condition of different heat inputs, Fig.4.29, shows typical martensitic microstructure and prior austenite grain boundaries surrounding martensite laths and decorated by precipitates. Large precipitates (most probably $M_{23}C_6$) form primarily on prior austenite boundaries and lath boundaries. Small precipitates form on the lath boundaries and inside the laths. Increased number of precipitates is clearly seen in the weld metal of the high heat input in the as-welded condition, Fig.4.29(c). This is due to the effect of slow cooling rate after welding as a result of high heat input which accelerates and allows enough time for nucleation and precipitation of carbides, and at the same time allows for more formation of lower carbon martensite than that formed in lower heat input conditions. The lath width increases with increasing heat input as shown in Fig. 4.29. The average lath width is about 2 microns. The reheating effect resulting from multilayer welding led to the formation of a microstructure similar to that expected for the HAZ (Fig.4.30). An increase in the measured ferrite volume fraction was found to occur with the increase of heat input, where the ferrite content measured by the ferrite scope was from 75%, to 77.5% and up to 82% respectively in the weld zone, table 4.10.

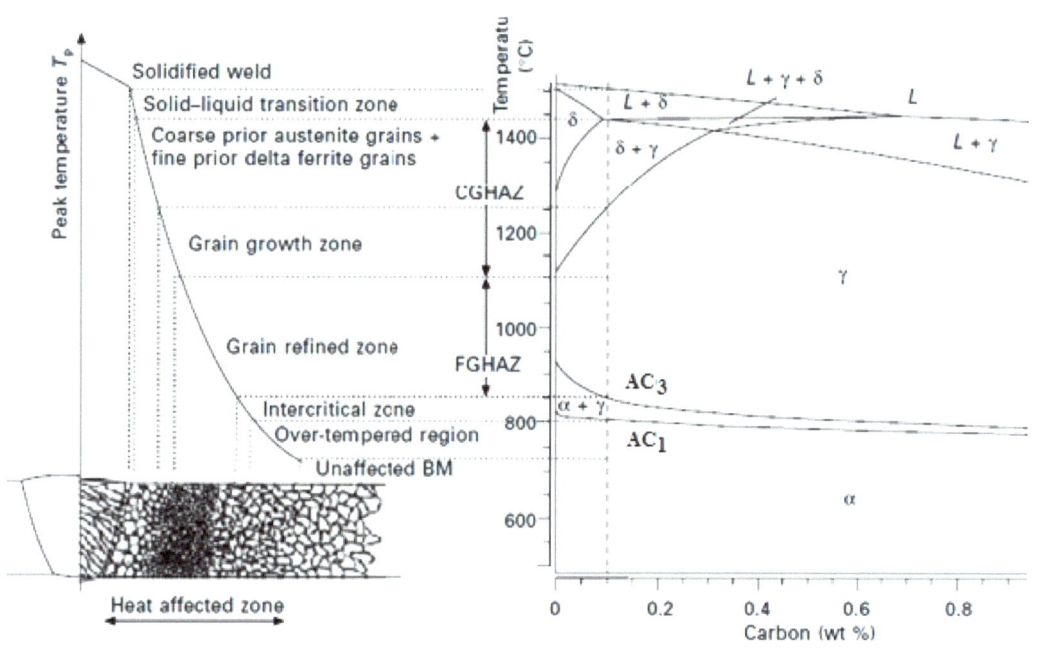

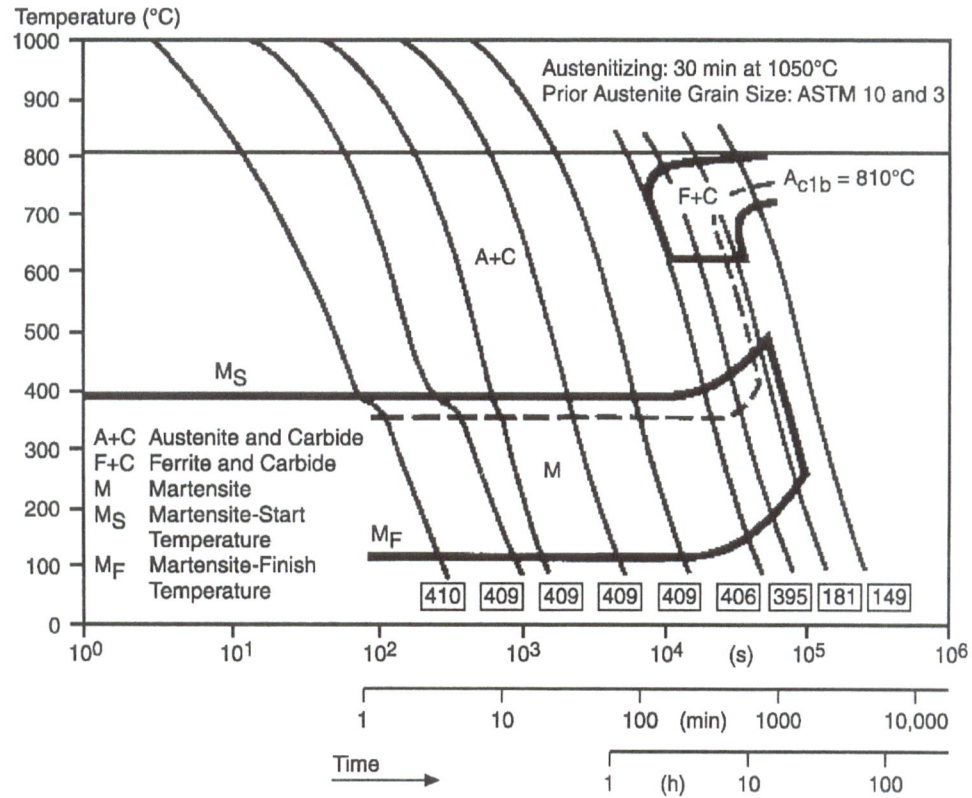

Fig.4.27. Schematic representations of microstructures developed in weld metal and HAZ as function of peak temperature during welding [94] and CCT diagram of steel P91.

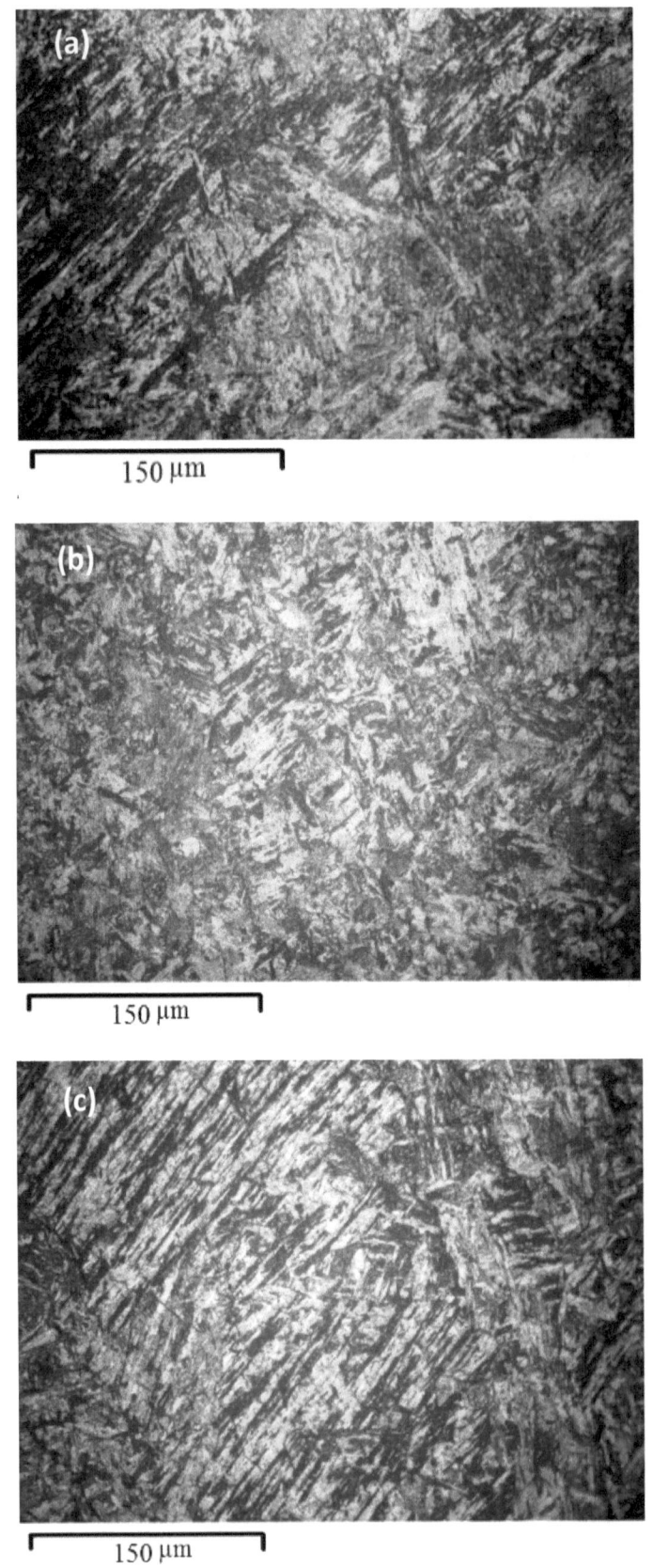

Fig. 4.28. Microstructure of weld metals of as-welded condition for different heat inputs showing fine un-tempered Martensite and acicular ferrite: a) L350, 75% ferrite (F); b) M350, 77.5% F; c) H350, 82% F.

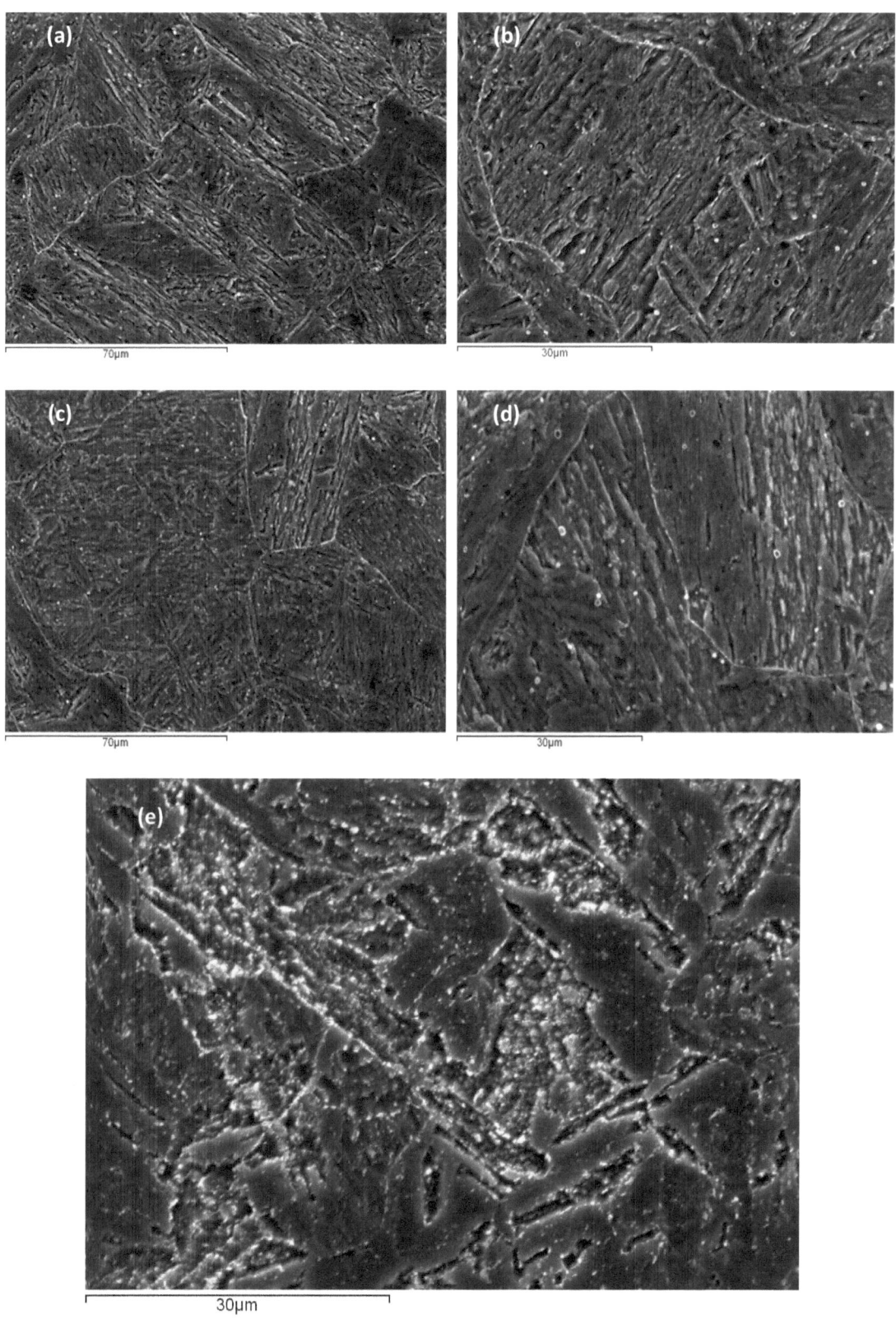

Fig. 4.29. SEM Microstructure of As-Welded condition Weld Metals: (a-b) L350. (c-d) M350. (e) H350.

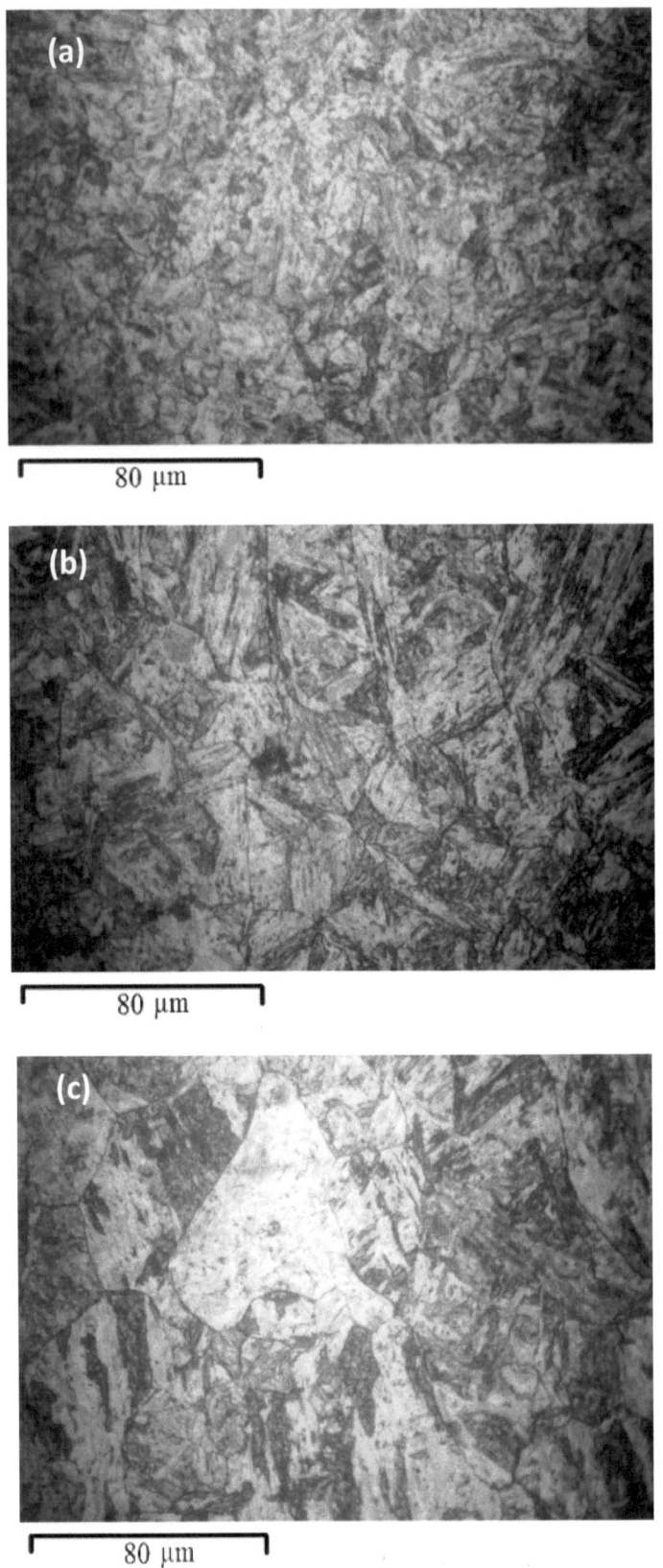

Fig. 4.30. Martensite microstructure at different heat affected areas between weld passes as a result of re-heat effect during multiple layers welding. a) Fine grains structure. b) Intermediate size grains structure. c) Coarse grains structure.

Table 4.10: Ferrite content measurement within Weld metals, HAZ and Base metals of different heat inputs/Heat Treatment conditions.

Sample Condition	Ave. Ferrite Cont.% At Weld	Ave. Ferrite Cont.% At HAZ	Ave. Ferrite Cont.% in Base Metal
L350	75.0	84.9	91.2
M350	77.2	86.8	
H350	82.2	90.9	
L760	95.0	94.9	94.0
M760	92.0	94.3	
H760	93.0	93.5	
L1050	94.6	90.0	93.9
M1050	94.4	90.5	
H1050	94.9	92.0	

Table 4.10 summarizes all ferrite content measurements of different conditions and at different location along the weldments for better comparison. As it was mentioned before the ferrite content is used as a measure of magnetic permeability and also as a measure of the degree of decomposition of martensite. When tempering is carried out the carbon content in martensite decreases due to decomposition of martensite and the precipitation of different carbides and carbonitrides which leads to increase in magnetic permeability and to increase in apparent ferrite content measured by the Ferritoscope. High heat input allows more precipitates of carbides and carbonitides to occur as seen from **Fig.4.29** resulting in less hard martensite with lower carbon content as was shown in the hardness distribution curves given in Fig. 4.1(a).

On the other hand, after subcritical PWHT a significant increase occurred in the measured ferrite volume fraction for all heat input conditions, where the ferrite content became 95%, 92% and 93% for low, medium, and high heat inputs respectively as shown in Table 4.10. This significant increase in the measured ferrite volume fraction came as a result of the decomposition of the un-tempered martensite during the subcritical PWHT **(Fig.4.31).** Some fine equiaxed ferrite grains are seen in the microstructure of **Fig 4.32**. In the case of the present study the tempering temperature (760 °C) was close to Ac_1 which explains the presence of the observed ferrite grains. On the other hand, the ferrite can also resulted due to the phase transformation came as a result of reheating effect of multiple passes welding which may reheat some areas within the weld to a temperature in the $\alpha + \gamma$ range, Fig.4.27.

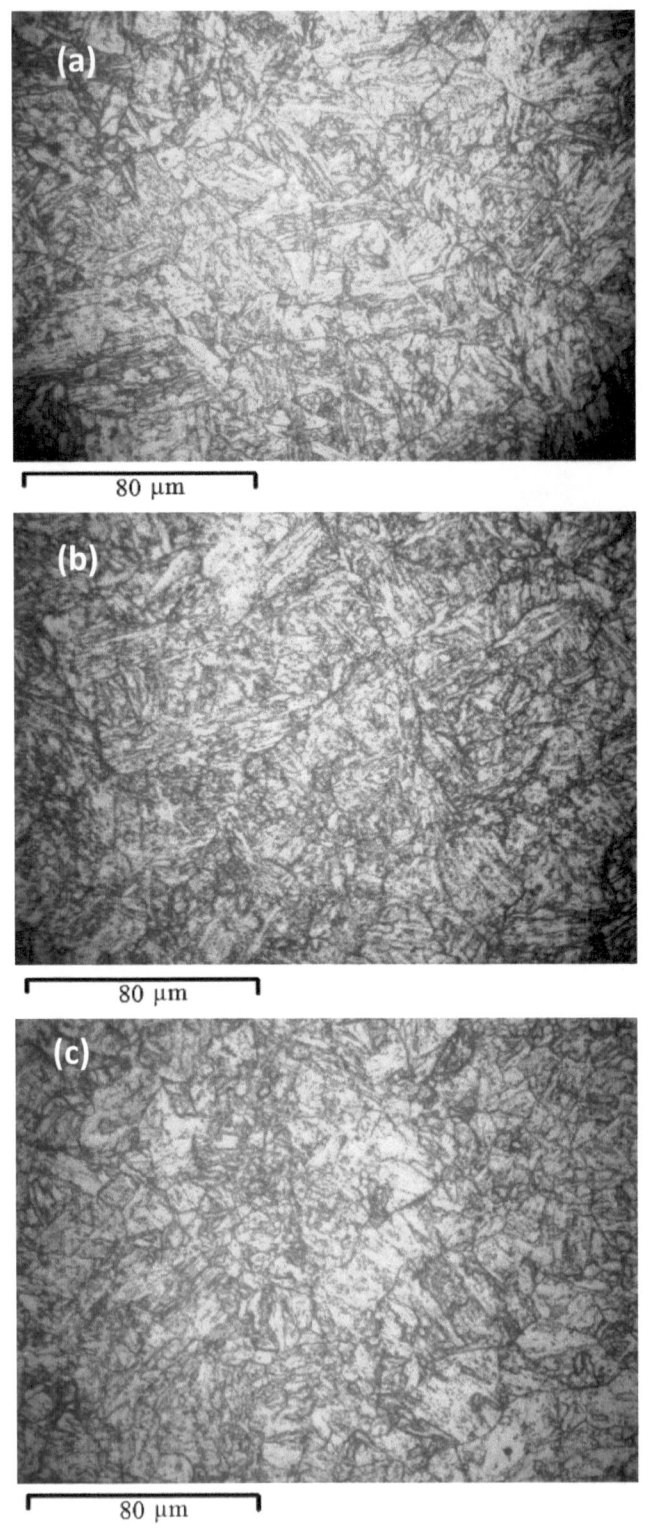

Fig. 4.31. Microstructure of weld metals after subcritical PWHT conditions:
a) L760, 95% ferrite (F); b) M760, 92% % F; c) H760, 93% F.

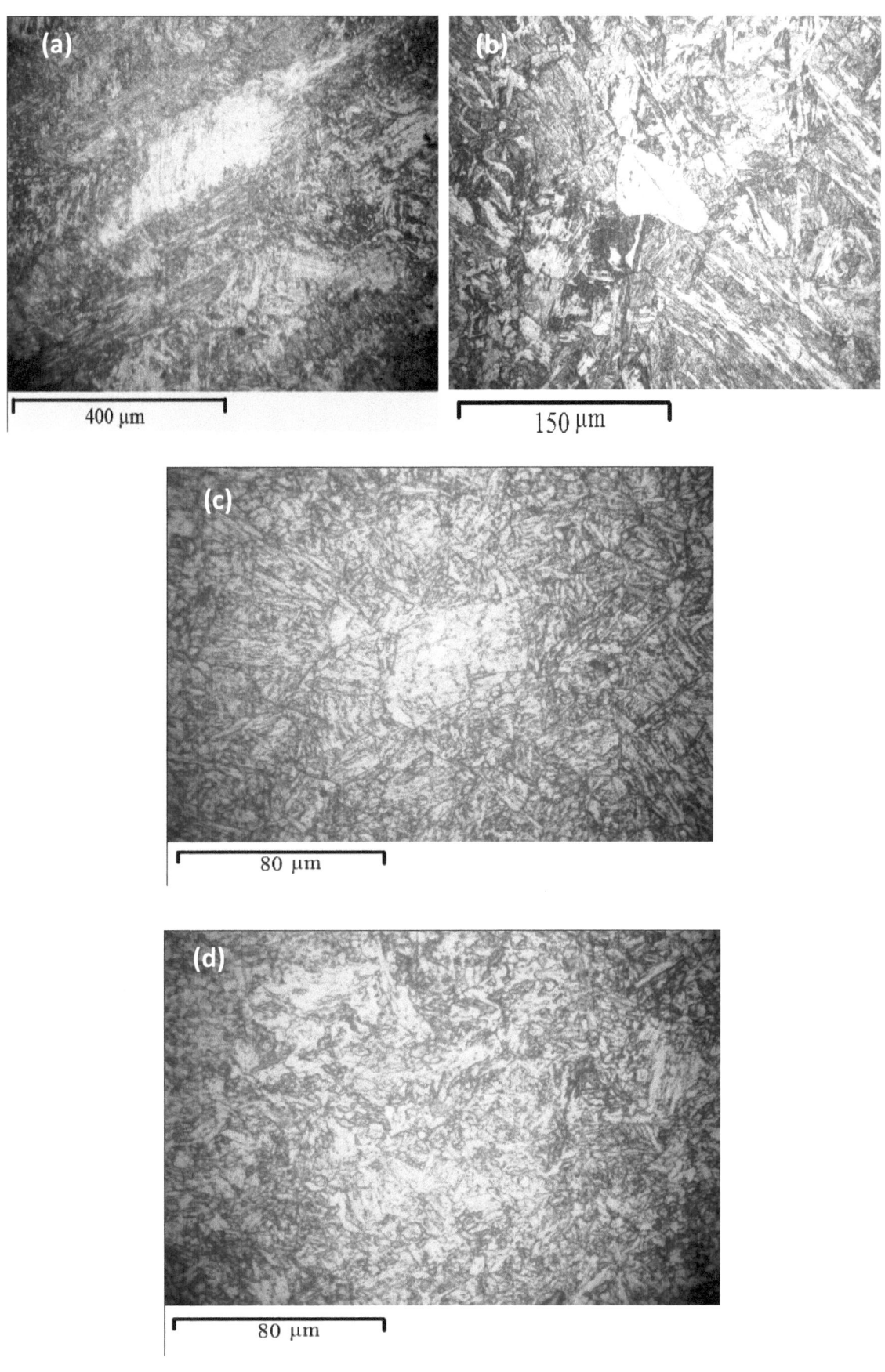

Fig. 4.32. Microstructure of weld metals: (a) M350. (b) L760, (c) M760, (d) H760. Some ferrite grains are seen in the microstructure.

The effect of heat input on measured ferrite contentment and hardness of the weld and heat affected zone is illustrated in Fig. **4.33 and 4.34.** From these figures it is clear that increasing heat input increases measured ferrite content (increases magnetic permeability) and decreases hardness.

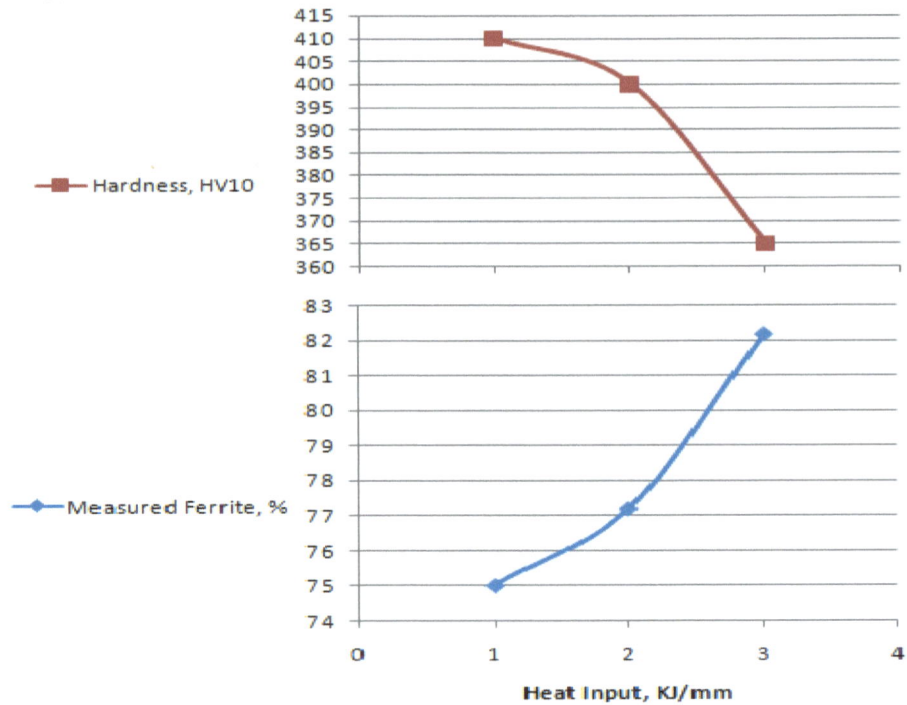

Fig.4.33 : Relation between measured ferrite content, resulted hardness and heat input for welds in as-welded condition.

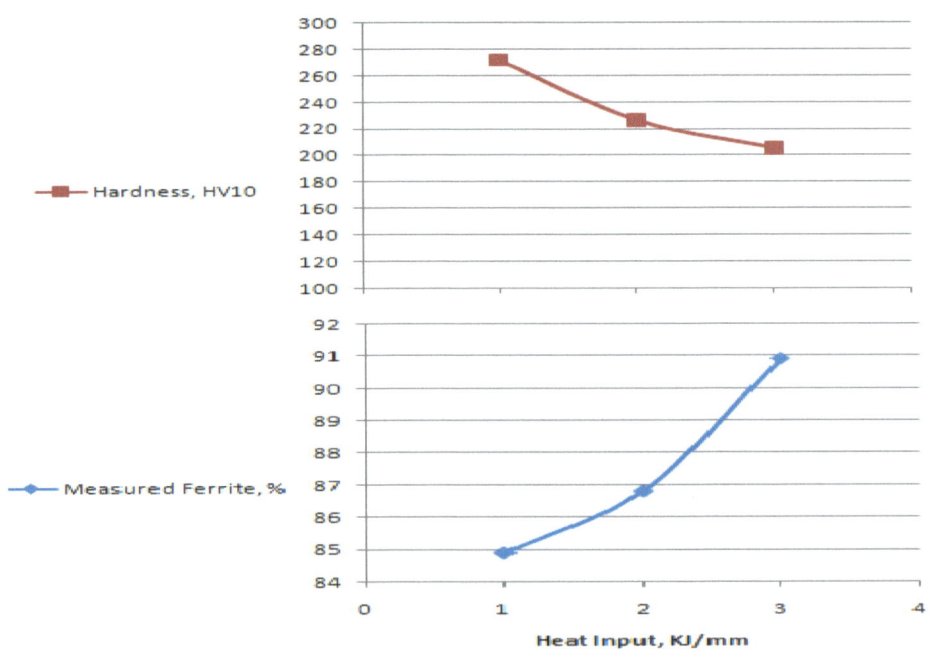

Fig. 4.34 : Relation between measured ferrite content, resulted hardness and heat input for HAZ in as-welded condition.

The correlation between measured ferrite content and hardness for all investigated conditions is given in Fig 4.35 from which it is clear that there is a good correlation with high correlation coefficient (0.85). Thus as the hardness decreases (due to tempering) the magnetic permeability increases and as a result the measured ferrite content increases. Actually the measured ferrite content is not a ferrite phase but tempered martensite with different degrees of decomposition of martensite.

Higher degree of microstructure homogeneity especially in the weld metal which was affected by the reheating effect of multi-pass welding was obtained for all heat input conditions after normalizing and tempering treatment as a result of phase transformations that occur during heating and cooling cycles of the normalization treatment (Fig.4.36). This microstructure homogeneity is also seen from the nearly equal measured ferrite content in Table 4.10 for the normalized and tempered conditions of the as welded metal welded by different heat inputs. Fig. 4.36(c) shows slightly finer microstructure resulting from the higher heat input condition compared to that obtained for both low and medium heat input conditions (Fig.4.36 (a-b)); and this with the observed higher percentage of precipitation may be the reason of the higher strength obtained in the high heat input normalized/tempered condition compared to that of low and medium heat input one (see Fig. 4.4(a)).

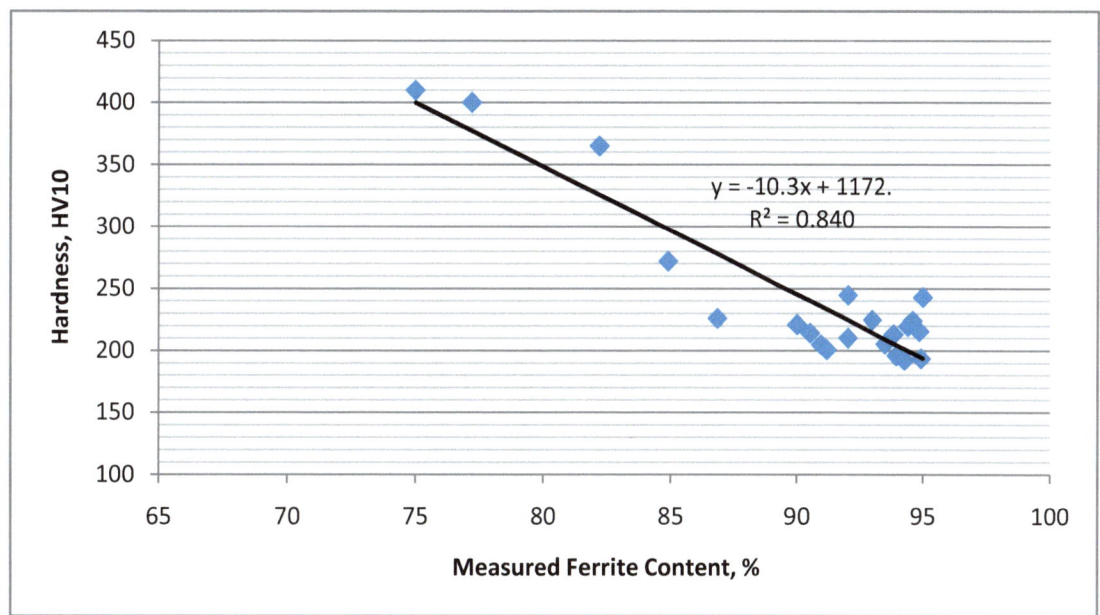

Fig 4.35 Relationship between measured ferrite content and hardness for as welded condition, tempered condition and normalized/tempered conditions.

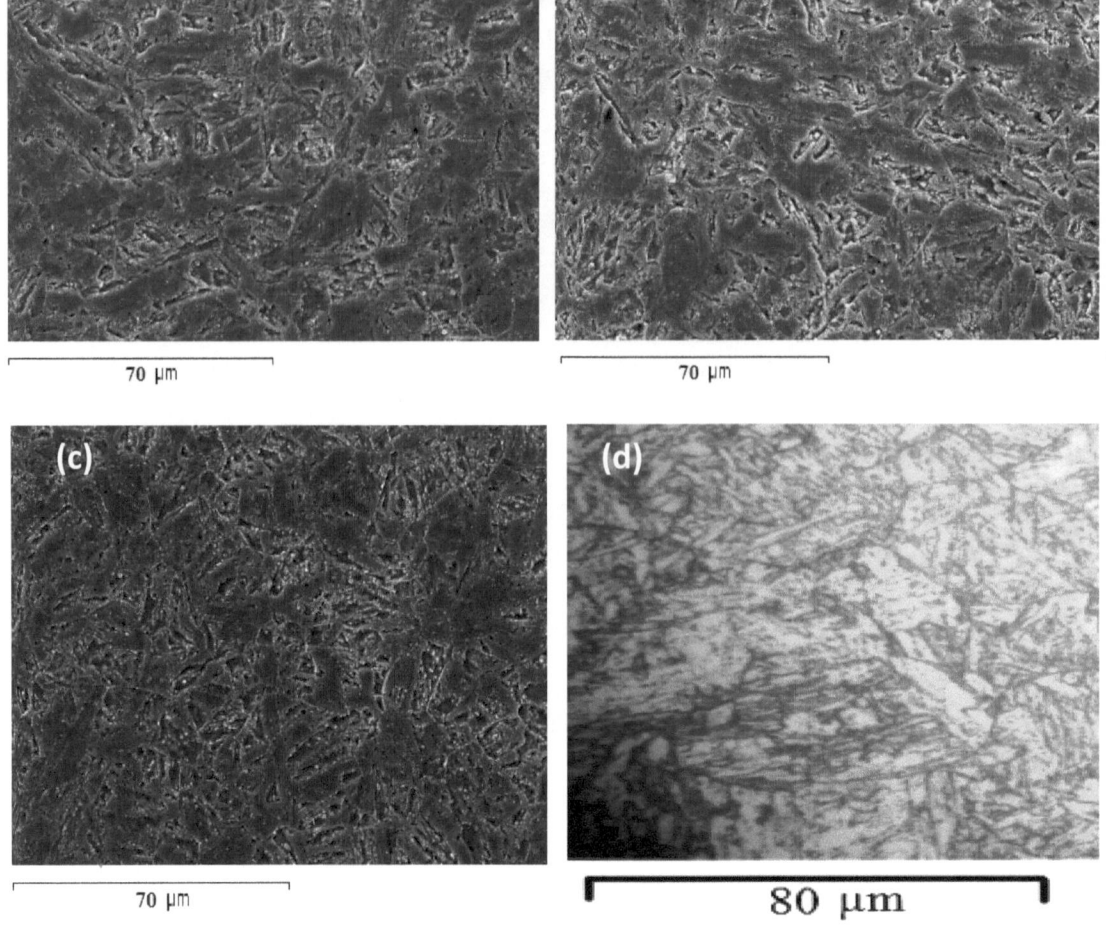

Fig. 4.36. SEM (a to c) Microstructure of Normalized/tempered Weld Metals:
a) L1050, 94.6% Ferrite (F). b) M1050, 94.4% F. c) H1050, 94.8% F.
d) H1050, 94.8% F light micrograph

X-ray diffraction patterns of the weld metals indicated similar patterns to that obtained from the as received base material (mainly martensitic structure) with no retained austenite (Fig.4.37).

Fig. 4.38 shows the resulted FWHM (Full width at half peak height) obtained from diffraction data of different weld metal conditions and the resulted crystallite sizes (Fig.4.39) obtained using the Scherrer Equation [96, 97]. The Scherrer analysis was conducted for the first peak only, because it has been reported [98] that better results are obtained from using diffraction peaks between 30 and 50 degree 2Θ. Micro-strains (like Lattice non-uniform distortion, dislocation and faults), welding residual stresses and crystallite size are important factors that affect the X-ray diffraction peak boarding. The FWHM obtained from the diffraction pattern analysis increase as micro-strain and residual stresses increase, and as the crystallite size decreases [98]. Fig. 4.38 illustrates the increase of FWHM with the increase of heat input. However, the slow cooling rate that takes place during high heat input welding allows enough time for proper arrangements of atoms in the lattice and this leads to minimum expected micro-strains and residual stresses. Thus it can be concluded from Fig. 4.39 that the main factor affecting the resulted FWHM is the crystallite size; which decreases as the heat input increase, and gives an idea about the high possibility of sub-grains formation enhanced by the slow cooling rate that allows dislocations re-arrangements in low-angle boundaries. However, welds of low and medium heat inputs/normalized condition have nearly similar strength as they also have similar crystallite sizes and apparent ferrite content (Fig. 4.4a & 4.39, Table 4.10).

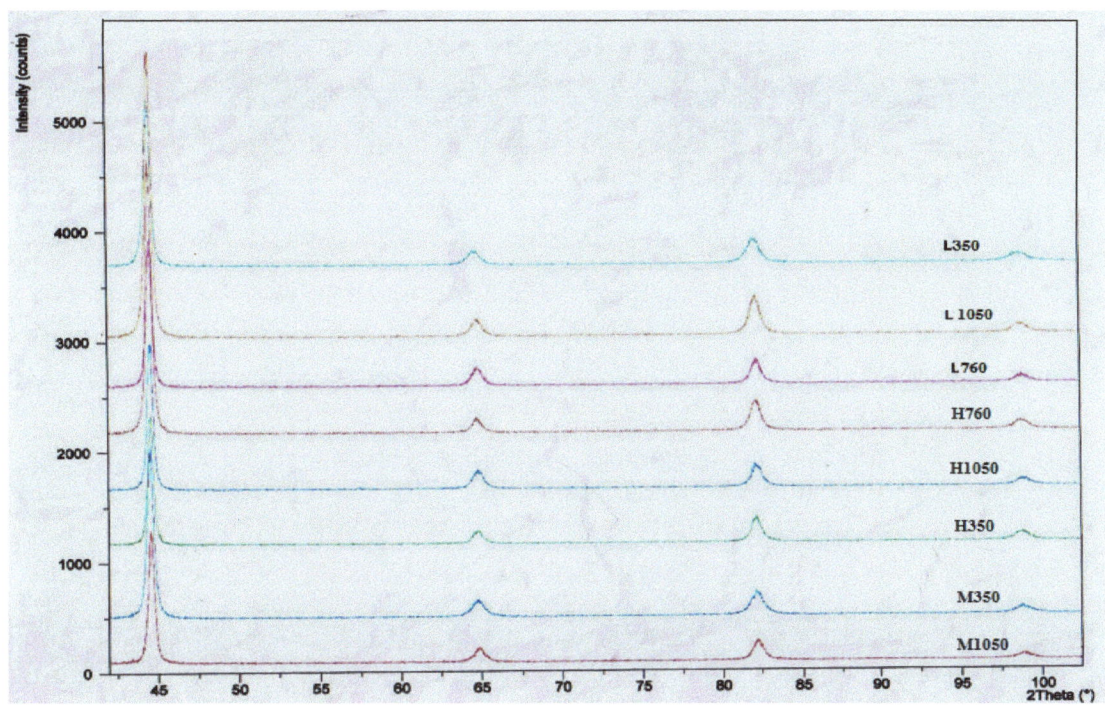

Fig. 4.37. X-ray diffraction pattern of base metal and weld metals of all heat/treatment conditions.

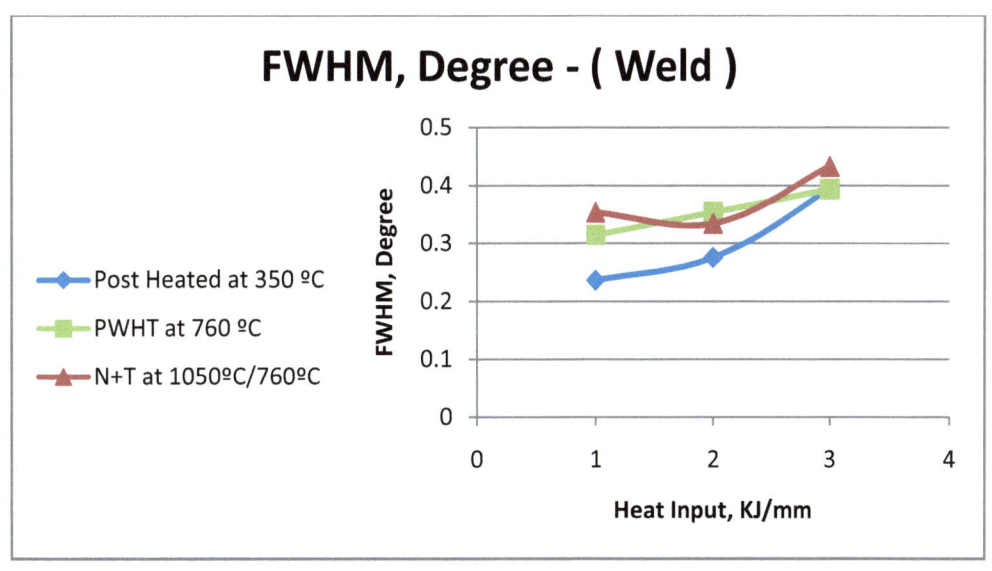

Fig. 4.38. FWHM resulted from X-diffraction analysis conducted on weld metals of different heat input/heat treatment conditions.

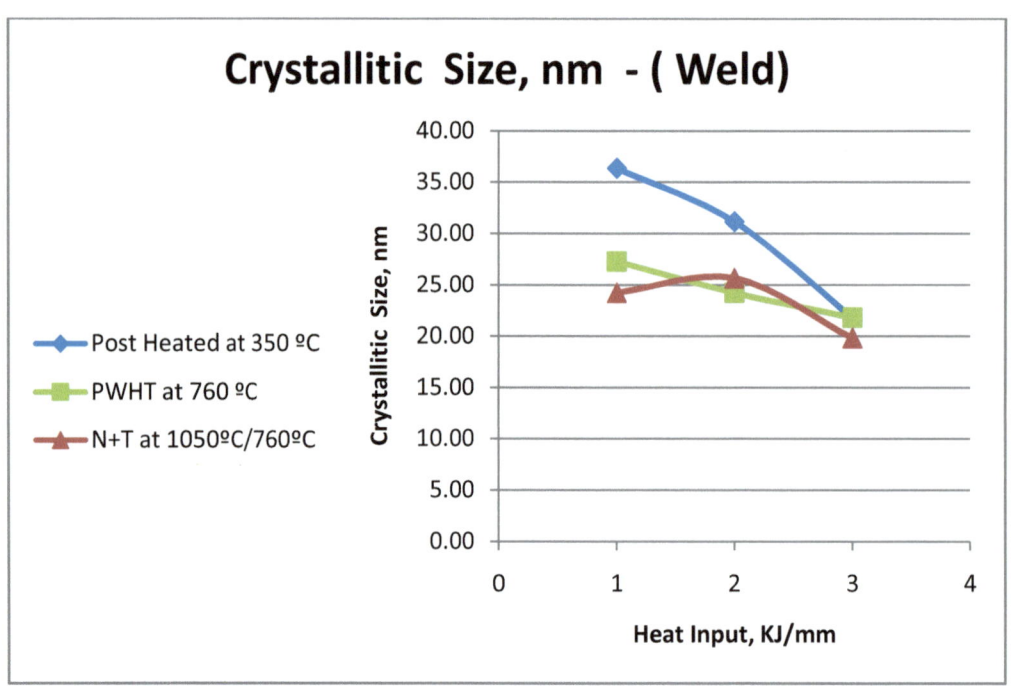

Fig. 4.39. Crystallite Size resulted from X-diffraction analysis conducted on weld metals of different heat input/heat treatment conditions.

A particularly important feature of interest in the P91 weld metal is the occurrence of patches of δ-ferrite. Fig. 4.40 shows different shapes of δ-ferrite within the weld metal for different heat inputs of the as-welded condition. It was observed that the volume fraction of δ-ferrite decreased with the increase of heat input due to lower cooling rates which allows time for phase transformation from δ-ferrite to austenite. The volume fraction of the δ-ferrite in the weld metal of the low, medium and high heat input of the as-welded conditions were found to be about 0.013%; 0.009% and 0.003 % respectively. The volume fractions of δ-ferrite was calculated as the ratio of the total area of δ-ferrite along the weld cross section to the total area of the weld cross section.

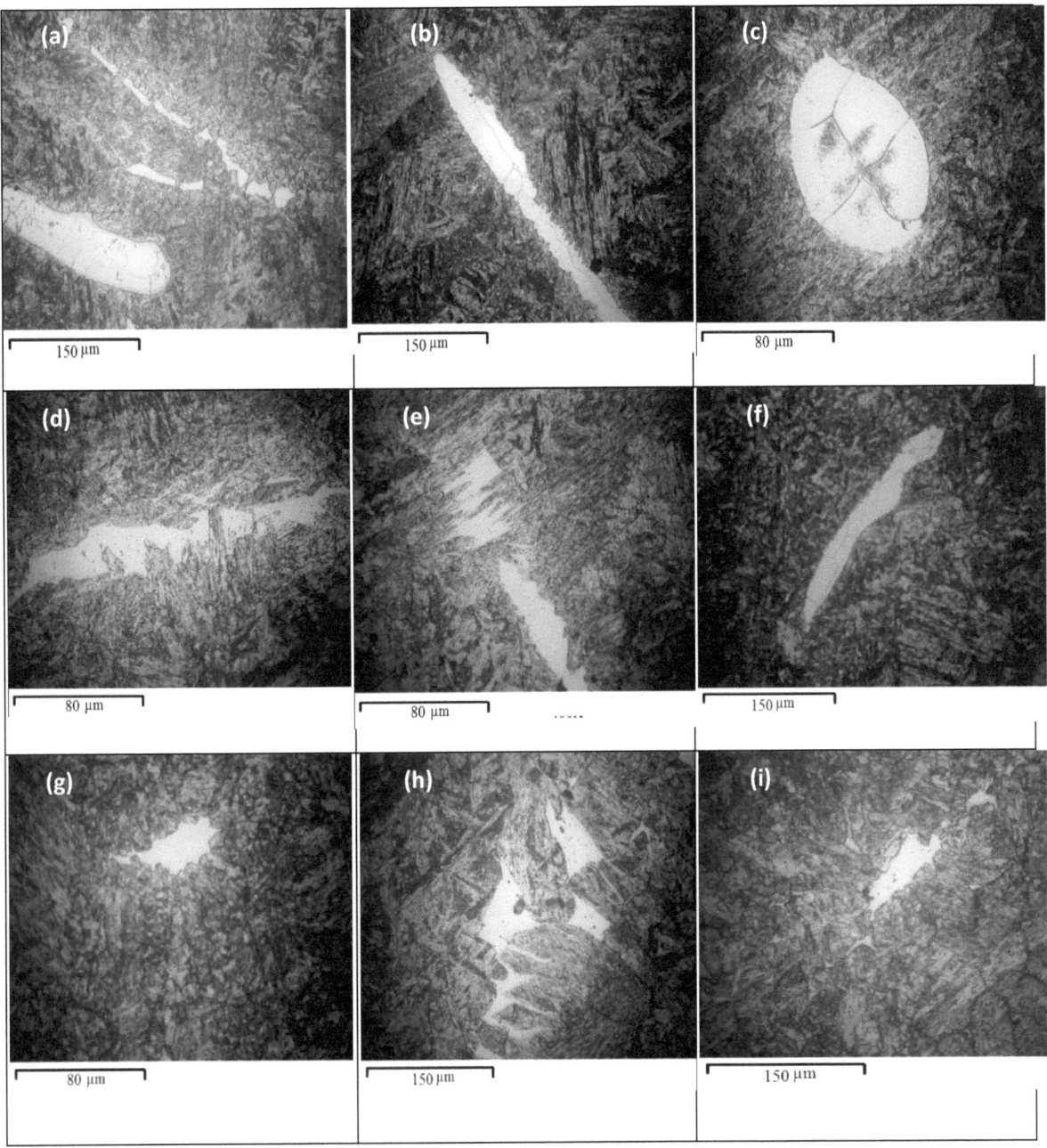

Fig. 4.40. Different types of formed δ-ferrite within the weld metals of as-welded conditions: (a-d) L350. (e, f) M350. (h, i) H350.

Arivazhagan [89] reported that the volume fraction of δ-ferrite increased with the increase of heat input in a low heat input range from 0.52 to 0.88 KJ/mm. In the present work the heat input values were in the range from 1 to 3 kJ/mm. Comparison of the results of Arivazhagan [89] with that obtained in this research leads to the conclusion that; volume fraction of δ-ferrite increases with the increase of heat input till a critical value of heat input and then decreases with the increase of heat input. This critical heat input value is still within the low heat input range (between 0.88 and 1.0 kJ/mm). It should be noted that, δ-ferrite remained in the microstructure after subcritical PWHT and after normalizing / tempering heat treatment (Fig.4.41 to 4.47). δ-ferrite also formed in the fusion area (Solid-liquid transition area) and in the course grain region in the heat affected zone (CGHAZ) located just beside the fusion line (Fig. 4.42).

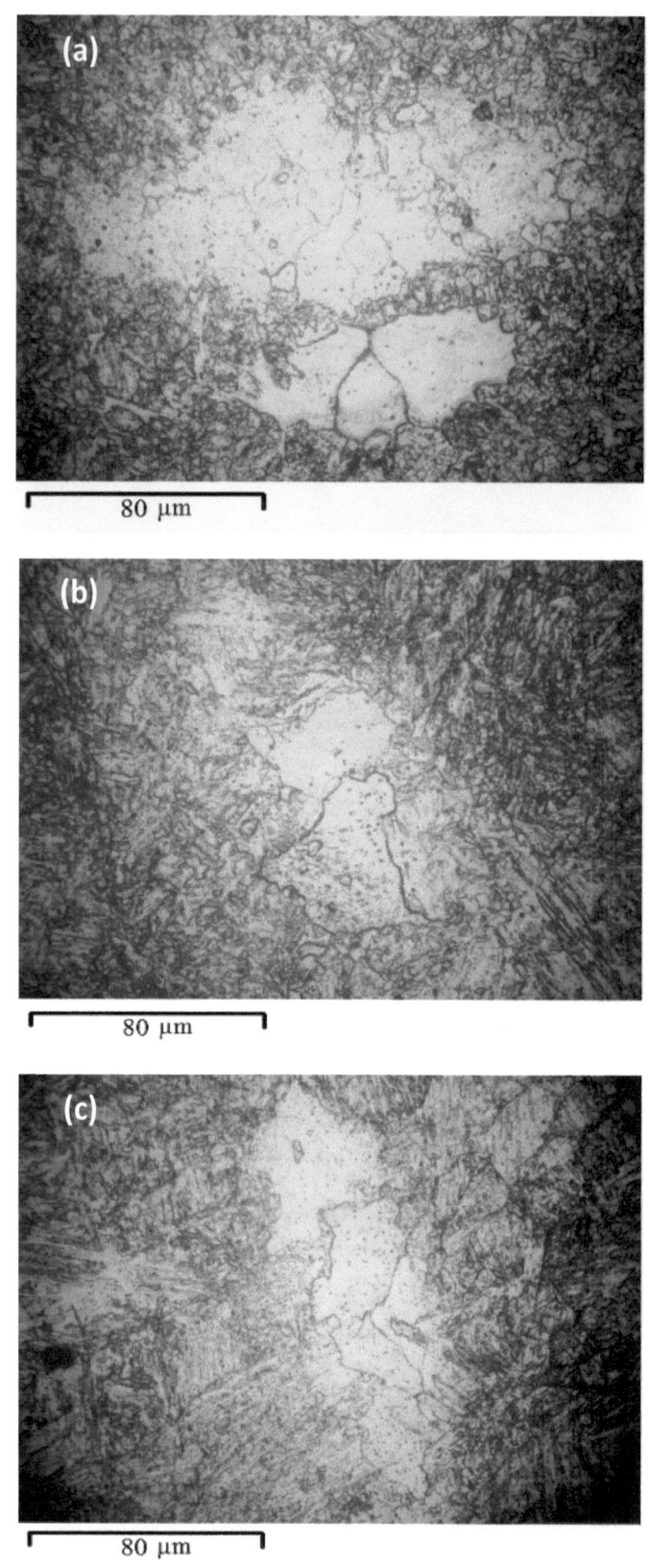

Fig. 4.41. δ-ferrite within the weld metal after subcritical PWHT: a) L760. b) M760. c) H760.

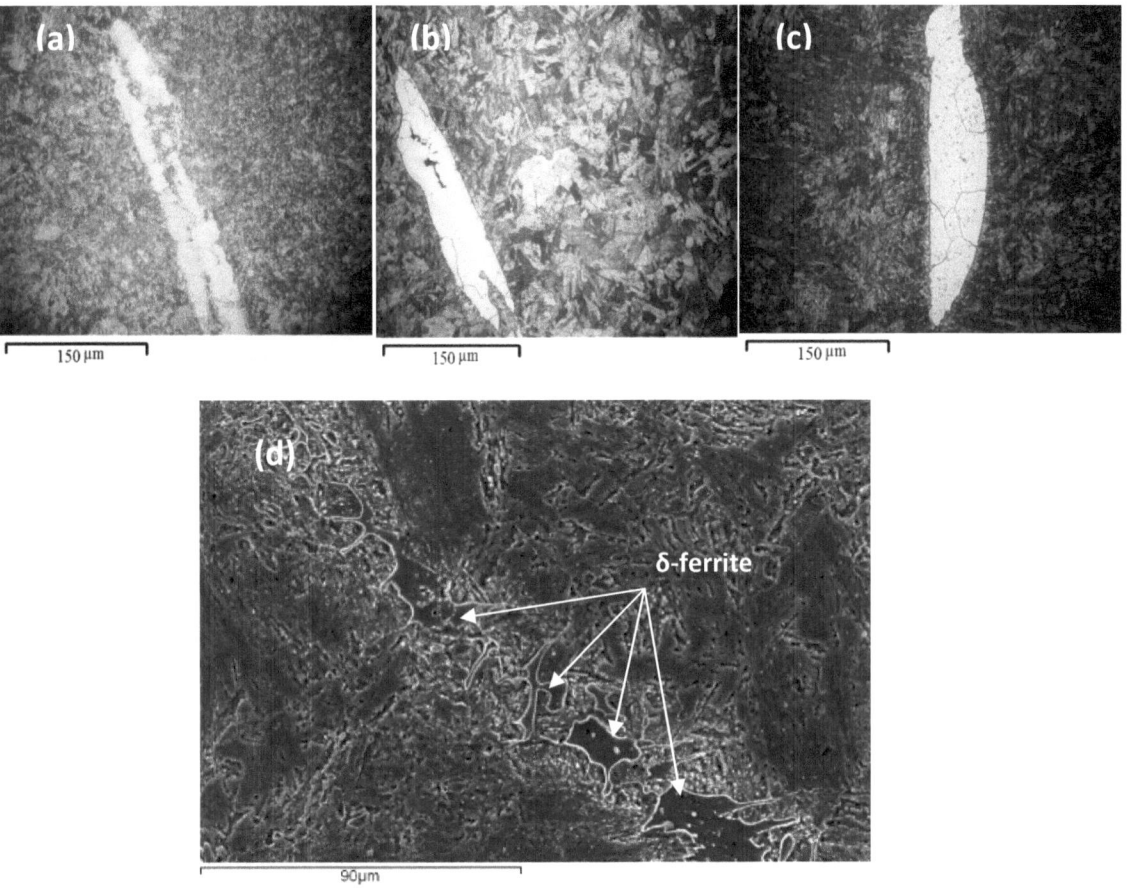

Fig. 4.42. δ-ferrite at fusion line and CGHAZ of different heat inputs of the as-welded and subcritical PWHT conditions: Optical micrographs; (a) L760. (b) M350. (C) H350. (d) SEM micrograph, fusion line, H760.

On the other hand, few δ-ferrite grains within the weld metal were found after normalizing and tempering treatment with a significant precipitation within the δ-ferrites grains, with the precipitations increasing with the increase of heat input (Fig.4.43). Low heat input showed the highest volume fraction of δ-ferrites grains.

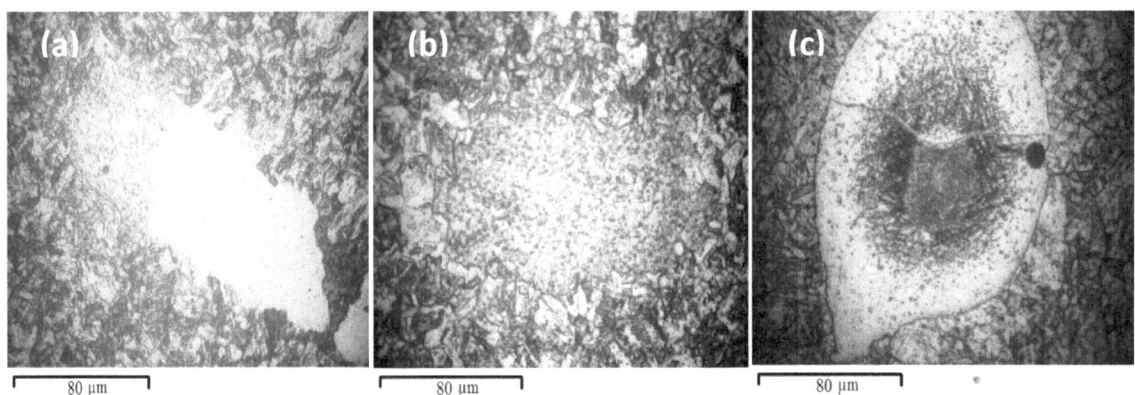

Fig.4.43. δ-ferrite within the weld metal after Normalizing and tempering treatment: a) L1050.   b) M1050.   c) H1050.

SEM and EDS analysis have been conducted on δ-ferrite, its grain, sub-grain boundaries, and precipitates observed within the matrix (Fig 4.44 to 4.47). SEM/ EDS analysis of weld metal of subcritical PWHT low heat input condition, Fig.4.44, shows δ-ferrites grain subdivided by sub-grains covered by coarse and high Cr/Mo content $M_{23}C_6$ carbides, and different needle like precipitates precipices within the matrix, which may be classified as MO_2C according to published data [79]. Also spherical precipitates were found which were classified as $M_{23}C_6$ based on the EDS results, Fig. 4.44(c) and Table 4.11. Fine ferrite grains with excessive $M_{23}C_6$ carbides precipitates along the grain boundary of δ-ferrites were also observed, Fig. 4.44(b) and Table 4.11. Same were observed through SEM/ EDS analysis for weld metal of subcritical PWHT medium heat input condition. Fig. 4.45 shows δ-ferrites with coarse high Cr/Mo content $M_{23}C_6$ carbides impeded within the matrix, Table 4.12. Fig. 4.46, for weld metal of subcritical PWHT high heat input condition shows δ-ferrite grain with fine $M_{23}C_6$ carbides impeded within the matrix, Fig. 4.46(c) and table 4.13. On the other hand, the ferrite grains along the δ-ferrite boundary showed less precipitation than that observed for weld metals of low and medium subcritical PWHT conditions, Fig. 4.46(b). This may be due to the slow cooling rate associated with high heat input which gives more time for diffusion of alloying elements out of these grains, and this may be the same reason of the observed low Cr and alloying elements content of the detected carbides within the δ-ferrites matrix (Fig. 4.46(c) and Table 4.13).

Another important observation on the effect of heat treatment on the δ-ferrite was detected through the EDS analysis of the δ-ferrite matrix in both subcritical and normalized/tempered conditions of the same heat input. Fig. 4.45(a) & table 4.12, shows higher percentage of alloying elements, especially, Cr and Mo than that obtained for the δ-ferrite in the normalized/tempered condition, Fig. 4.47 & table 4.14, which came as a result of improved homogeneity resulted from phase transformation effect through austenization process. However, this effect may be also the reason for the formation of low Cr/Mo fine $M_{23}C_6$ within the δ-ferrite matrix and the absence of carbides in the matrix of the ferrite grains along the δ-ferrite boundary in the same heat input condition (Medium heat input) after normalizing/tempering treatment, Fig. 4.47 & table 4.14, compared to that of subcritical medium heat input condition, Fig. 4.45 & table 4.12.

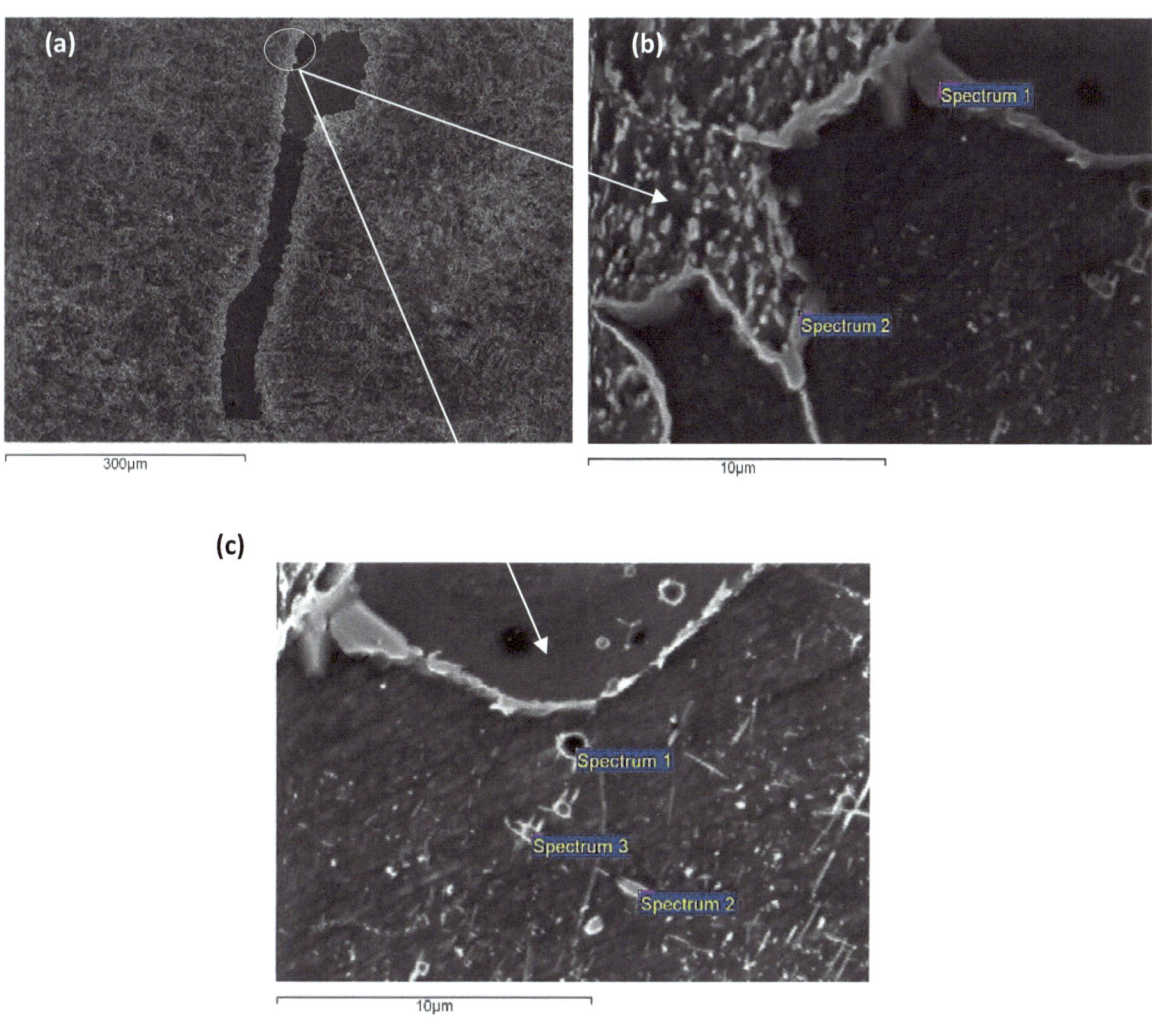

Fig. 4.44. SEM Microstructure of Delta Ferrite in the Weld Metal of L760 and location of EDS analysis conducted (Table 4.10). (a) Delta Ferrite with sub-grains structure. (b) Fine Ferrite grains with excessive precipitates at the boundary of delta ferrite. (c) Sub-grain inside delta ferrite grain with different types of precipitates.

Table 4.11: Results of EDS analysis related to precipitates indicated through Fig.4.44, wt %.

Element	C	Si	V	Cr	Mn	Fe	Ni	Nb	Mo
Spec.1b	16.55	0.26	0.31	19.64	0.45	59.85	-----	0.23	2.73
Spec.2b	20.06	0.38	0.32	17.68	0.64	57.95	0.55	0.26	2.15
Spec.1c	25.67	0.91	0.33	13.87	0.63	55.04	0.43	0.08	0.66
Spec.2c	27.69	0.35	0.39	18.41	0.57	51.12	0.52	----	0.95
Spec.3c	34.00	0.50	0.20	11.17	0.71	53.46	0.32	----	0.66

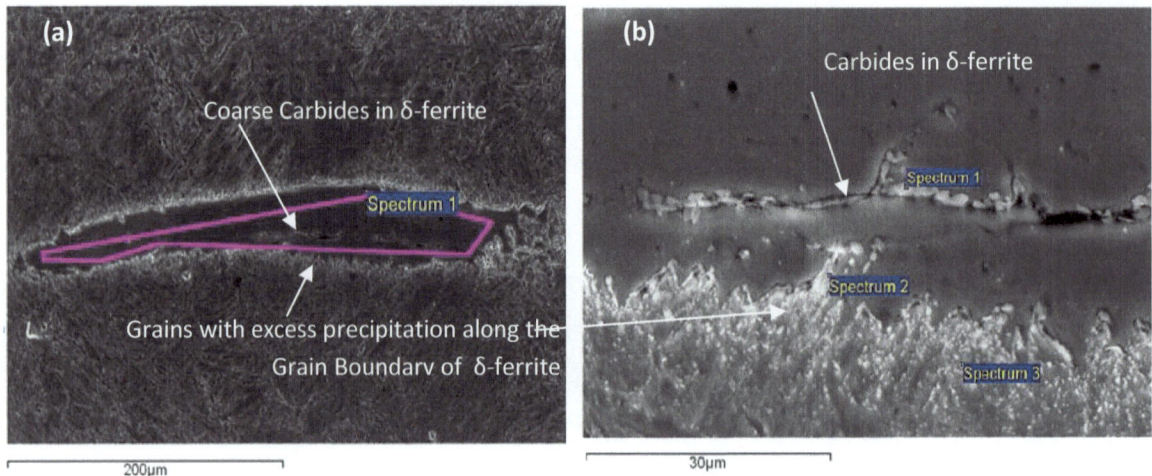

Fig. 4.45. SEM Microstructure of δ-ferrite in the Weld Metal of M760 and location of EDS analysis conducted (Table 4.11). (a) δ-ferrite. (b) Grains with excessive precipitates along the boundary of δ-ferrite.

Table 4.12: Results of EDS analysis related to precipitates indicated through Fig.4.45, wt %.

Element	C	Si	V	Cr	Mn	Fe	Ni	Nb	Mo	Other
Spec.1a (δ-ferrite)	3.76	0.55	0.25	17.97	0.88	73.43	0.57	0.39	2.20	----
Spec.1b	12.3	0.43	0.27	26.39	1.07	50.39	0.53	----	5.07	W 0.29 Ti 0.02
Spec.2b	16.6	0.27	0.29	22.49	0.85	52.38	0.43	---	3.53	Ti 0.06
Spec.3b	19.3	0.32	0.27	12.37	0.75	62.55	0.41	0.11	1.36	----

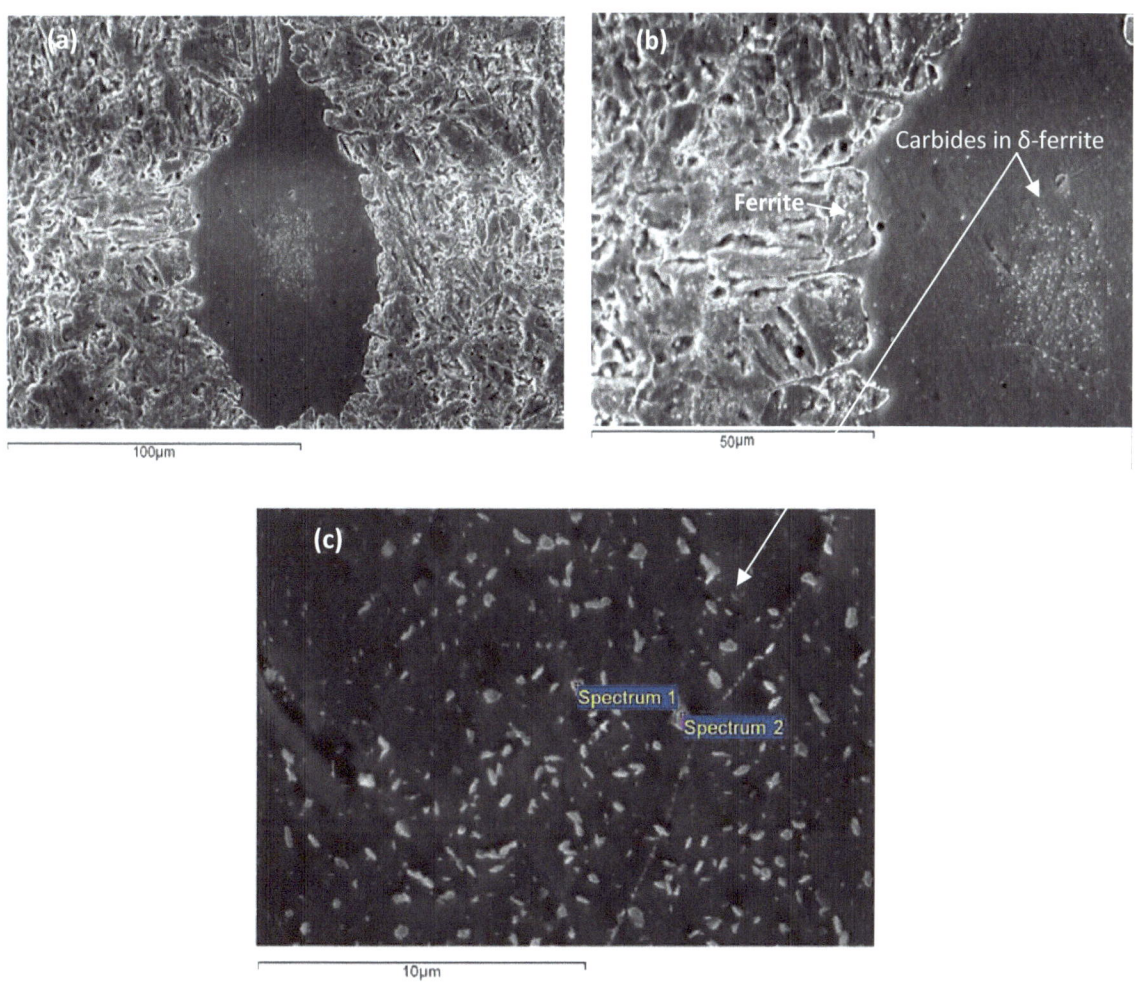

Fig. 4.46. SEM Microstructure of δ-ferrite in the Weld Metal of H760 and location of EDS analysis conducted (Table 4.13).

Table 4.13: Results of EDS analysis related to precipitates indicated through Fig.4.46, wt %.

Element	C	Si	V	Cr	Mn	Fe	Ni	Nb	Mo	Other
Spec.1c	41.5	0.07	----	5.77	0.37	49.94	0.24	----	2.07	----
Spec.2c	37.1	----	0.15	6.36	1.13	51.94	0.29	---	3.59	W 0.16

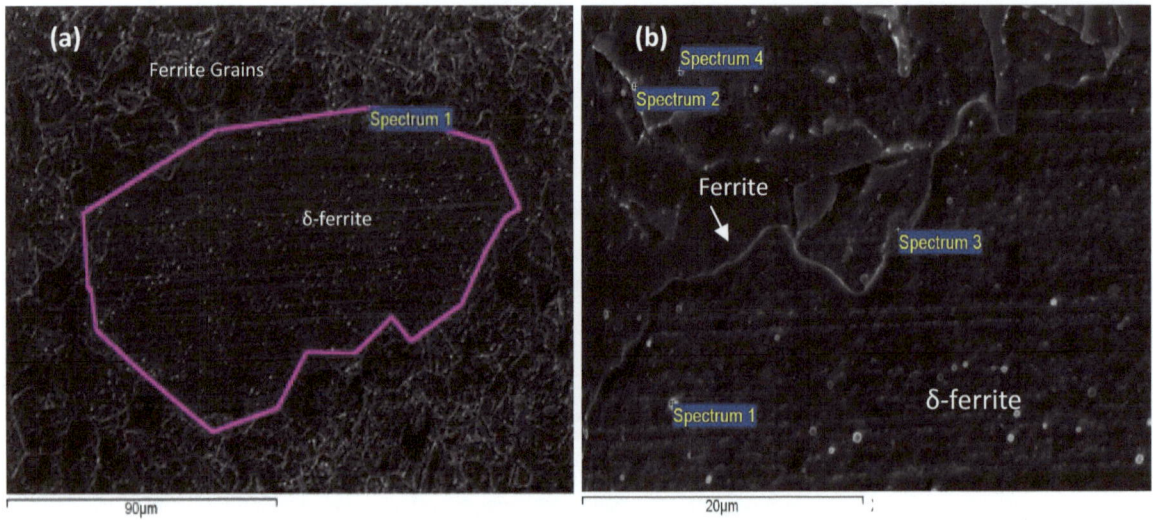

Fig.4.47. SEM Microstructure of δ-ferrite in the Weld Metal of M1050 and Locations of conducted EDS analysis. (a) δ-ferrite with fine distributed carbides along the matrix. (b) Ferrite with low precipitation percentage along the δ-ferrite boundary

Table 4.14: Results of EDS analysis related to precipitates indicated through Fig.4.47, wt %.

Element	C	Si	V	Cr	Mn	Fe	Ni	Nb	Mo	Other
Spec.1a (δ-ferrite)	3.27	0.12	0.28	7.49	0.58	86.5	0.03	0.24	1.11	Ti 0.04 W 0.33
Spec.1b	25.8	0.06	0.23	5.98	0.57	67.03	0.18	----	-----	W 0.14
Spec.2b	29.5	0.19	0.23	5.83	0.64	62.02	0.23	0.28	1.06	----
Spec.3b	29.0	0.09	0.20	5.59	0.44	63.67	0.16	0.08	0.69	Ti 0.04
Spec.4b	28.9	0.37	0.30	5.81	0.58	63.38	0.10	----	0.48	----

The following different microstructures were identified along the HAZ toward the base metal in both as-welded and subcritical PWHT conditions, Fig. 4.27:

i) Coarse grain region (CGHAZ): Area near the fusion boundary that reaches a temperature well above Ac_3 during welding; where, Ac_3 is the temperature at which transformation of ferrite into austenite is completed upon heating. Carbides which constitute the main obstacle to growth of the austenite grains dissolve resulting in coarse grains of austenite. This austenite transform into high carbon and hard martensite on cooling (Fig.4.48(a)).

ii) Fine grain region (FGHAZ): Away from the fusion boundary where the peak temperature (Tp) is lower, but still above Ac_3. Austenite grain growth is limited by lower temperature and by the incomplete dissolution of carbides. Fine grained austenite is produced, which subsequently transforms into low carbon slit softer martensite (Fig.4.48(b)).

iii) Inter critical region (ICHAZ): Here the peak temperature Tp is lower than Ac_3 but higher than Ac_1; where Ac_1 is the temperature at which transformation of ferrite into austenite is started on heating, and this results in partial reversion to austenite on heating. The new austenite nucleates at the prior austenite grain boundaries and the martensite lath boundaries; whereas the remainder of the microstructure is simply tempered. The austenite transforms into un-tempered martensite on cooling (Fig.4.48(c)).

iv) Over tempered region: With Tp below Ac_1 the original microstructure of the plate material undergoes further tempering (Fig.4.48(d-e)).

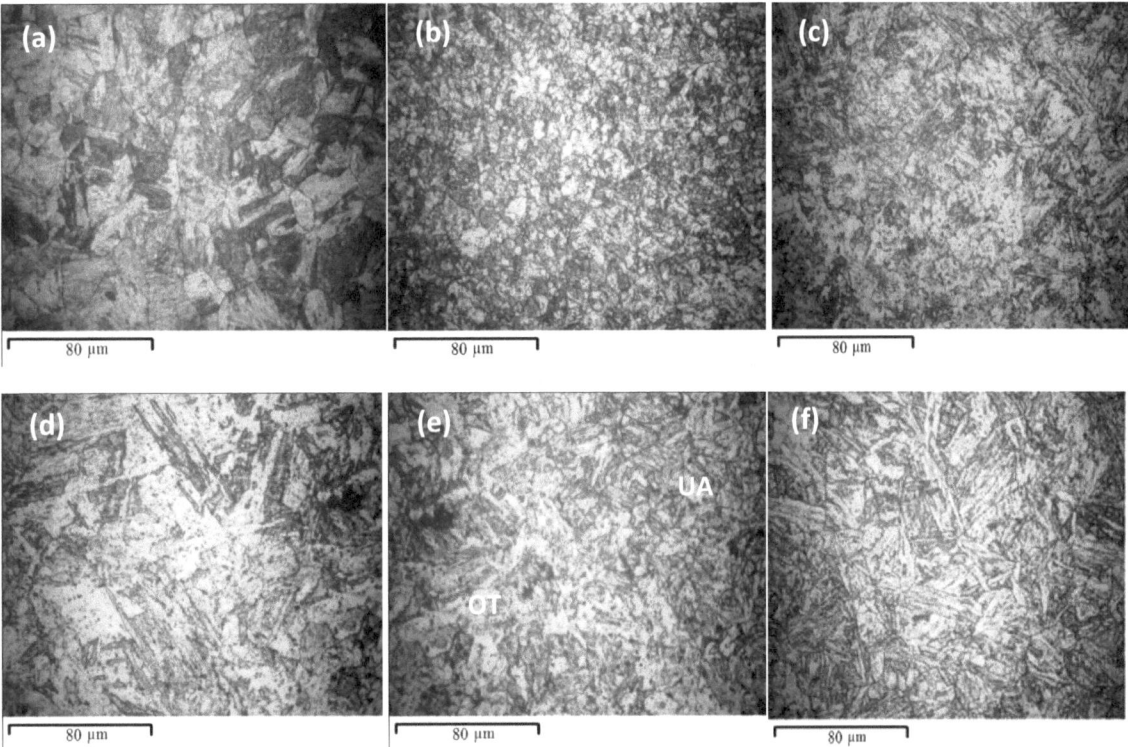

Fig. 4.48. Different areas at the HAZ of low heat input (L350): (a) CGHAZ. (b) FGHAZ e. (c) ICHAZ. (d) Over tempering zone (OT). (e) Boundary between over tempered (OT) and un-affected base metal (UA). (f) Un-affected Base Metal.

In fact, the obtained microstructures within the weld and HAZ depend strongly on the obtained peak temperature and the cooling rate. However, the maximum obtained peak temperature at welding pool, its gradient effect, cooling rate and the width of formed HAZ are all directly dependent on the heat input used during the welding cycle; the higher the heat input, the higher the peak temperature, the lower the cooling rate, and the larger the temperature gradient effect and the wider the HAZ [95].

To sum up, the higher the heat input:

- The lower the cooling rate of weld pool leading to the formation of lower volume fraction of δ-ferrite.
- The coarser the CGHAZ grains and the harder the martensite formed after cooling as a result of complete dissolution of carbides within the austenite grains.
- The wider the CGHAZ, FGHAZ, ICHAZ and over tempered HAZ.

Measured ferrite content increased in the HAZ area as a result of martensite decomposition during subcritical PWHT which minimize at the same time the differences between measured ferrite contents of different heat inputs, 94.9%, 94.3% and 93.5% respectively. On the other hand, normalizing and tempering treatment minimize the differences between the HAZ and weld metal structures for the same heat input (low and medium) as a result of phase transformations during heating and cooling cycles of normalizing treatment. However, the differences in microstructures of weld metal, HAZ, and base metal did not disappear completely after normalizing/tempering heat treatment for the high heat input (Fig. 4.49 and 4.50). Average ferrite content measurements conducted on the HAZ area of the three heat inputs after normalizing and tempering treatment shows slight differences in measured ferrite content between low and medium heat inputs compared to that of the high heat input condition, 90%, 90.5% and 92% respectively. However, X-ray diffraction patterns of the HAZs indicated similar patterns to that obtained from the as received base material and weld metals of different conditions (mainly martensitic structure) with no retained austenite.

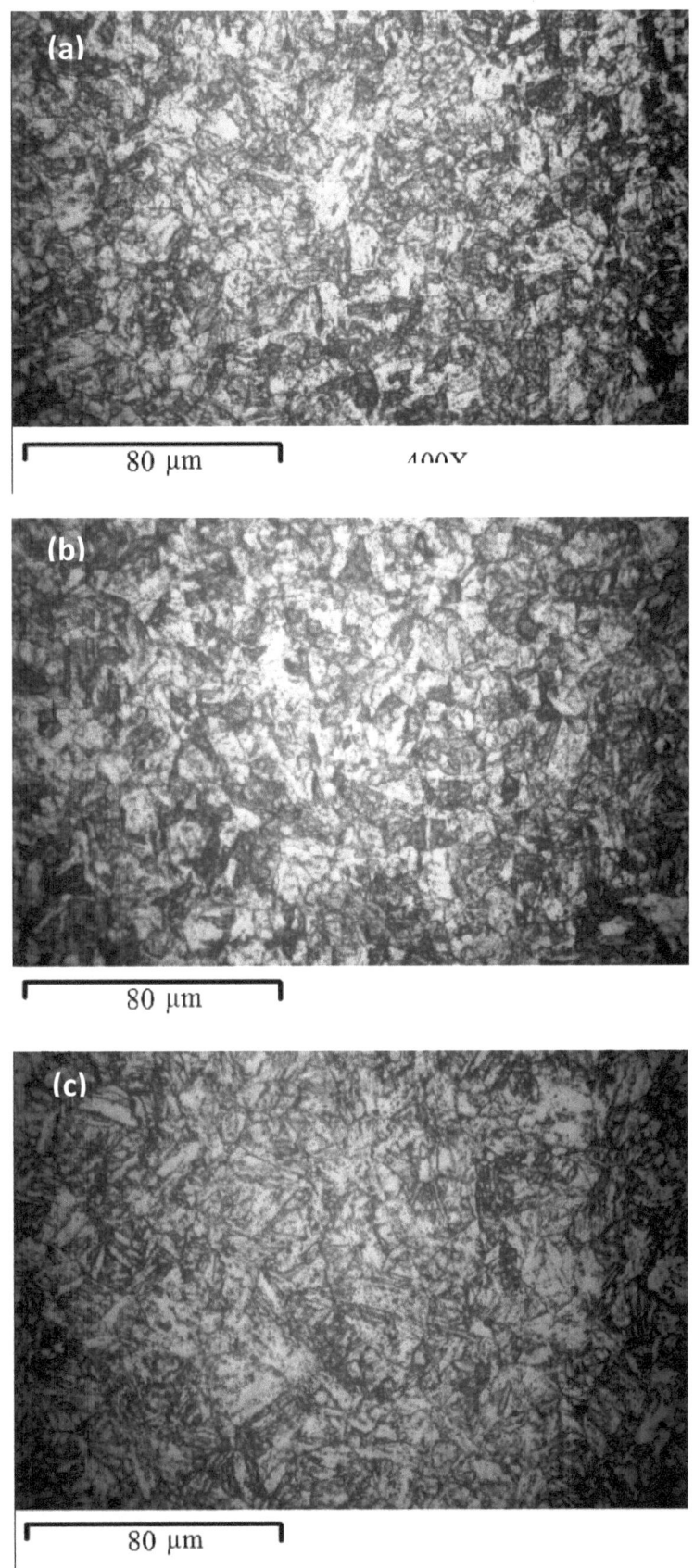

Fig. 4.49. Optical Micrographs of weld metals of normalized/tempered conditions: (a) L1050. (b) M1050. (C) H1050.

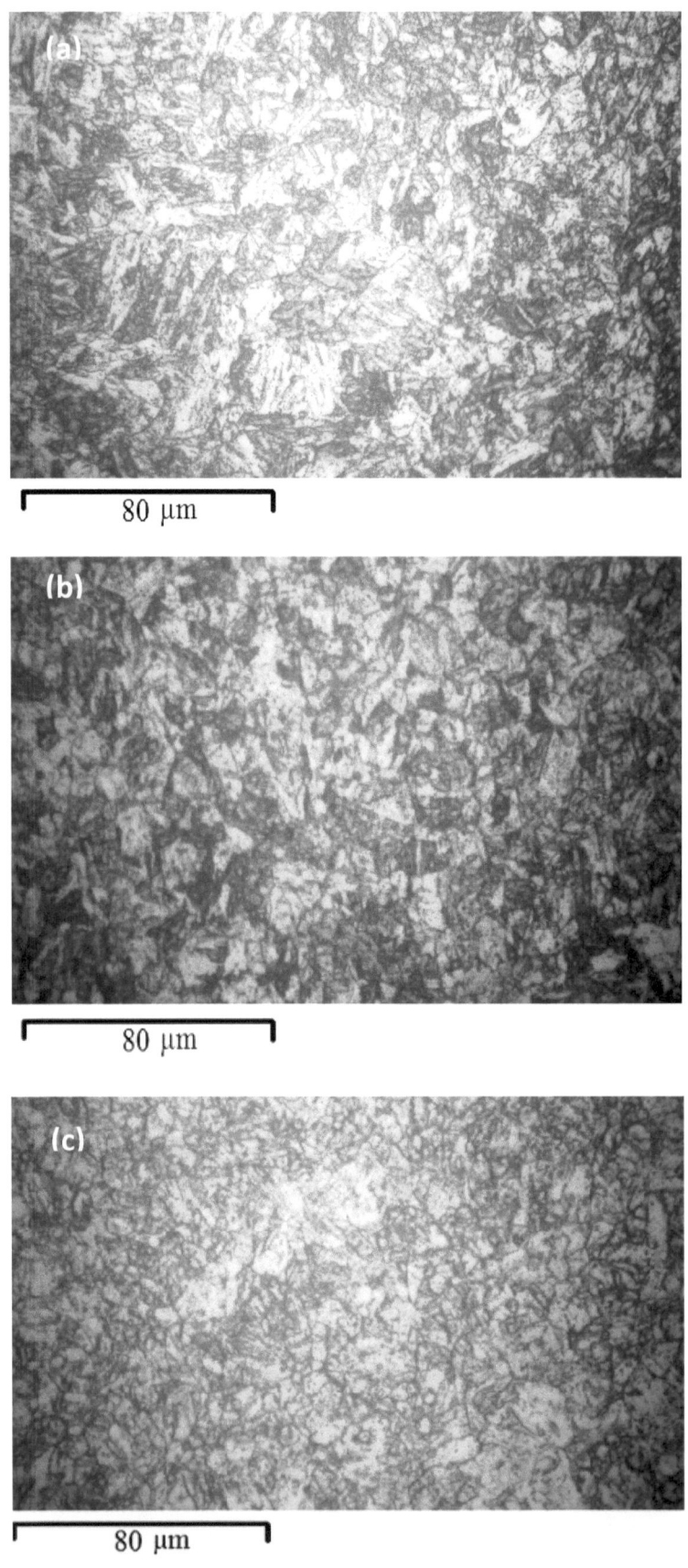

Fig. 4.50. Optical Micrograph of HAZ after N&T treatment: (a) L1050. (b) M1050. (c) H1050.

Another particularly important feature of interest is the possibility of bainite-like microstructure as a result of martensite decomposition during multiple tempering at 760°C. A percentage of bainitic ferrite was detected at base materials and HAZ/Base boundaries in the subcritical PWHT conditions (Fig.4.50). The presence of bainitic ferrite is more evident in the high heat input condition which may indicate an increase of the probability of bainite formation with the increase of heat input. In fact, the continuous cooling transformation (CCT) diagram published for P91 material didn't indicate the probability of bainitic transformation **[99, 100]**. However, Mile demonstrated that in P91 steel a microstructure which appeared to be of bainitic character could be the case if multiple heat treatments during a manufacturing process (e.g. bending) took place **[101]**. Milović reported the formation of small amounts of bainite in P91 steels during cooling from austenization temperature as the martensite formation was preceded by separation of a smaller quantity of bainite **[102]**. He reported also the formation of bainite in addition to martensite in simulated HAZs using simulation temperatures up to 1150°C and with cooling time $t_{8/5}$ of 40 sec. followed by PWHT at 730°C **[102]**.

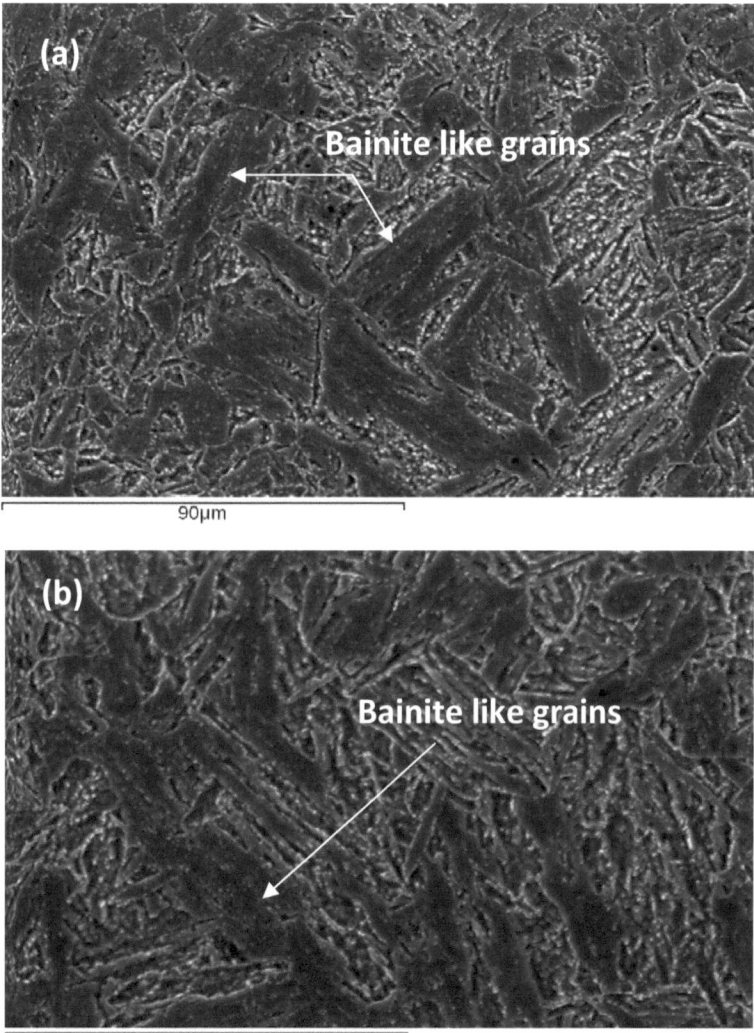

Fig. 4.51. SEM showing bainite like grains in the subcritical PWHT condition. (a) HAZ area near base metal of high heat input condition H760. (b) Base metal affected by double tempering at 760 °C.

Summary and Conclusions

1. Heat input and heat treatment were found to have a great influence on the obtained microstructure of P91 weldments which have direct effect on the mechanical properties and weldments serviceability.
2. Tempered martensite were the main observed phase through different areas of P91 weldments. Bainite and δ-ferrite can be formed under certain conditions.
3. δ-ferrite was observed in the weld metal at all heat input/treatment conditions, and its volume fraction increased with increase of heat input till a critical value lower than 1.15 KJ/mm, then decreased with the increase of heat input.
4. δ-ferrite was observed at the fusion boundary and CGHAZ of all as-welded and subcritical PWHT conditions, but was not reported for normalized and tempered conditions.
5. The formation of a soft zone formed in the HAZ portion subjected to over tempering effect and may act as the source of earlier creep failure during service.
6. Normalizing and tempering treatment reduces the amount of retained δ-ferrite within the weld metal, and results in the optimum degree of microstructure homogeneity along the weldments of all heat input conditions as a result of phase transformation during normalizing treatment.
7. For both subcritical post weld heat treatment, and normalizing and tempering heat treatment conditions the heat input range from 1 to 2.28 KJ/mm yields the optimum properties.
8. The normalized/tempered heat treatment show a significant increase in the stress rupture life mounting to about 90% more than that of the subcritical PWHT.
9. Stress rupture life decrease with the increase of heat input for both subcritical and normalized/tempered conditions.
10. The stress corrosion cracking tests conducted on all conditions illustrates good resistance of this material and its weldements to the stress corrosion attack whatever the heat input or heat treatment is.
11. Pitting resistance of P91 weldements decreases with the increase of heat input, but normalized/tempered conditions shows better resistance than that of sub-critical PWHT.
12. Weld metal in all heat input/heat treatment conditions has better pitting resistance than HAZ and base metal.